DIGITAL
BIOLOGY

How nature is transforming our technology and our lives

PETER J. BENTLEY

SIMON & SCHUSTER

NEW YORK   LONDON   TORONTO   SYDNEY   SINGAPORE

SIMON & SCHUSTER
Rockefeller Center
1230 Avenue of the Americas
New York, NY 10020

Previously published in Great Britain in 2001 by Headline Book Publishing

SIMON & SCHUSTER and colophon are registered trademarks
of Simon & Schuster, Inc.

For information regarding special discounts for bulk purchases,
please contact Simon & Schuster Special Sales at 1-800-456-6798 or
business@simonandschuster.com

Designed by Jan Pisciotta

Manufactured in the United States of America

10 9 8 7 6 5 4 3 2 1

Library of Congress Cataloging-in-Publication Data
Bentley, P. J.
   Digital biology: how nature is transforming our technology and our lives/Peter J. Bentley.
       p.   cm.
   Includes bibliographical references (p.).
   1. Biological systems—Computer simulation. 2. Adaptive control systems. I. Title.
QH324.2.B46   2002
570'.1'13—dc21   2001054987
ISBN 0-7432-0447-6

Remember our bet, Tina?

CONTENTS

1

INTRODUCTION

Imagine a future world where computers can create universes—digital environments made from binary ones and zeros. Imagine that within these universes there exist biological forms that reproduce, grow, and think. Imagine plantlike forms, ant colonies, immune systems, and brains, all adapting, evolving, and getting better at solving problems. Imagine if our computers became greenhouses for a new kind of nature. Just think what digital biology could do for us. Perhaps it could evolve new designs for us, think up ways to detect fraud using digital neurons, or solve scheduling problems with ants. Perhaps it could detect hackers with immune systems or create music from the patterns of growth of digital seashells. Perhaps it would allow our computers to become creative and inventive.

Now stop imagining.

We are all becoming blasé about computers. The whirring, chittering box sitting on (or underneath) your table is probably no longer an off-putting device for you. More likely, you regard it as a necessary evil. When working, it becomes part of the furniture, and when not working, it becomes something to hurl abuse at. You cannot even escape it when leaving your office. Everywhere you go, you hear people talking in a new language of e-mails, dot-com domain names, and template files. What existed only in the laboratories of computer scientists ten years ago now forms part of everyday conversation for the least computer literate of us. Computers and computer software are everywhere, and yet, except for a few of the "nerdy types," they are largely ignored or taken for granted by us. Somehow this seems unfair. When some of the most exciting and inspiring new developments of our technology are happening inside the minds of these benign cream-colored boxes, to disregard them is to ignore something wonderful. So I shall not ignore them. I shall do the opposite: I will focus on them. By doing so, I will change the way you think about computers.

Follow me into a different universe—the digital universe of our computers. I will show you the marvels that inhabit this strange new environment. You'll find them familiar, but different. Alike, but diverse. The digital entities I want you to meet are not incomprehensible collections of numbers or equations. They are just the same as you and the natural world that surrounds you. They may live and die within digital domains, but they are every bit as biological as you. Together, they comprise *digital biology*.

To make this journey, we must abandon our physical forms and don digital bodies, for we cannot exist in a digital universe as we are. Ready? Here we go.

THROUGH THE LOOKING SCREEN

The journey from our universe to the digital universe is instantaneous. We open our digital eyes and see . . . trees. But these trees are not static. As we watch, we see them grow—from a small seedling to a vast, towering pine. The immense and beautiful complexity of the branches develops before our eyes; leaves form in symmetrical patterns; stems thicken into trunks. A forest grows up around us, the trees all trying to outreach their neighbors in the quest for the sunlight. At our feet are other plants—ivy sending out tendrils to the trees and then swarming up them like leafy snakes. Ferns unfurl themselves, and we see their intricate fractal-like forms spread wide to catch the low level of light at the forest floor. We move through the forest and spot some seashells on the ground. Again, they grow before our eyes, the spirals and patterns emerging like ripples on water. When you pick up one of the shells and put it to your ear, instead of hearing the sea, you hear the pattern of the growing shell, transformed into music. These are not ordinary plants and shells. They are not made from atoms and molecules, and they are not growing in our universe. They are digital plants, made from the flow of electrons within the digital universe of a computer.

Let me slow down time in this digital universe. The growth of the trees and plants slows and stops. Now other movement becomes evident. On the ground we see insects scurrying along. A long line of ants marches forward, each following the pheromone trails left by its companions ahead of it. As we watch, the first few ants reach a specific point on the ground and turn, heading toward a new location, with the ants behind them following like an over-ambitious conga line. As we stand back, we can see that the ants are tracing out a complex and zigzagging route across the ground. Looking up, we see the sky filled with a flock of birds, flying in astonishing formations. They circle around each other, dive up and down, and suddenly all change direction at the same time— yet never does a single bird collide with another. But there are still many more digital marvels in this universe for me to show you. Let me adjust our size.

We shrink rapidly, until the ants tower above us, as high as skyscrapers. Suddenly everything becomes dark. We have been swallowed by some digital creature. Don't be alarmed—I arranged for this to happen. I would like to show you around inside.

We watch the flow of cells from within the veins of the creature. Most are completely normal, but once in a while, we spot a cluster of cells stuck together. In the center of the cluster is a virus or some other unwanted intruder into this creature. It is surrounded by large, colorless cells, which are attacking it. These are the leukocytes—the white blood cells of our digital creature that detect anything that is not regarded as "normal." We move into the bone marrow of the creature. It is here that new white blood cells are constructed. Using the patterns laid down in the DNA of this creature, new cells are created constantly, each configured to attack a different invader. As we watch, a new cell is created and released into the bloodstream. If this cell finds an intruder, it will immediately clone itself, increasing the number of cells designed to attack this particular type of unwanted guest. Throughout the network of veins, an ever-changing police force of white blood cells patrols the creature, ensuring that immunity to all types of attacker is maintained at all times.

We move again inside the digital creature and find ourselves in the brain. Around us is a dazzling electrical storm of activity. Interconnected neurons fire electrical signals at each other in a vastly complicated network, informing their companions to stop or start firing themselves. In parallel, chemical signals are constantly emitted from the cells, and new connections between cells are grown or lost. The unceasing activity around us embraces electrical signals generated by the creature's senses and emits a never-ending stream of signals to the muscles and organs of the creature, causing it to react to its environment and learn, plan, and predict things in its world.

I shrink us again in the digital universe. We are now so small that we are within a single cell of the creature. In fact, we can see molecules within the cell. One vast spiraling molecule dominates the view—a strand of digital DNA held within the nucleus of every cell of the creature. The DNA defines how the creature is grown from a single cell. It also specifies how the creature should mature and controls the production of cells for the immune system. The separate genes in the DNA are rules, designed to turn on and off the production of proteins. The proteins trigger new cells to grow, cause them to differentiate into different types of cell, and even tell cells to die. The DNA of this creature defines a complex recipe of actions and counteractions, safeguards and repair mechanisms.

I grow us back to our original size; the creature we were in shuffles away. Again trees, ants, and birds surround us. Digital DNA is held within all of these aspects of digital biology, defining the growth and behavior of everything we see. But to show you how the genes of the digital DNA are created, I must speed up time in this digital universe.

As the movement of the birds and insects disappears in a blur of activity, once again we see the trees and plants growing around us. I continue to speed up the passage of time, until even the growth of trees becomes too fast to see properly. Now all we can see is a nebulous landscape of digital biology, with forms appearing and disappearing as their lives are lived in split seconds. And as these biological forms live, as they find their partners and reproduce, they form part of an ever-changing environment. Whether tree or ant, if the offspring inherit some genes that allow them to perform a little better in the digital universe, then they will survive a little longer and on average will have a few more offspring than the others. So the more useful genes become more numerous in the populations, resulting in more successful creatures. This continuous process of change is known as evolution, and it shapes digital biology in the digital universes of our computers just as it shapes the natural world in our own universe. Indeed, as we watch, new types of tree form, changing generation by generation. New types of ant, with subtly different behaviors, emerge, more complex brains develop, more efficient immune systems grow. But our time grows short, so we must depart from the digital universe and return to more familiar surroundings.

NATURAL TECHNOLOGIES

Now that we have left the digital universe, I must come clean. Digital biology does not exist in a single universe, in a single computer. All of the aspects of digital biology I have just described exist in different, isolated digital universes. The digital ants will never meet the flock of birds, nor will they crawl on the ferns in the digital forest. Even the immune system and brain of the creature that swallowed us do not coexist in the same digital universe. But the fact that they do exist is undeniable.

I am one of hundreds of scientists who spend their time understanding the processes of nature and enabling those same processes to happen in computers. To achieve this, we create simple digital universes in computers, using laws of physics laid down in our software. Within these digital universes, we grow a new type of nature. We have harnessed the power of natural processes such as evolution and growth. Digital embryos grow from digital DNA, digital plants evolve, ant colonies swarm, neural network brains learn. We use such digital biology to *evolve* solutions to problems, such as methods for detecting fraud. We explore how immune systems can be created within computer networks and used to attack hackers. We discover how to use colonies of computational ants to search for better solutions to scheduling problems. We examine how architectural designs can be grown from a set of digital genes into adult form. We find out how to use digital neural networks to detect the difference between benign and malignant cancer cells. We learn how to develop colonies of digital cells that have the behavior of fire. By using the natural processes responsible for life within computer software, we are overturning all preconceptions of what computers can and cannot do.

Through our work, we find new and highly efficient ways of solving today's problems. We also learn about the techniques we have borrowed from nature. We find out more about our own origins, about the mechanisms behind evolution and embryology. We learn how plants grow, how animals develop, how our immune systems work, and even how we think.

These new techniques will form the next generation of our technology. By understanding the solutions of nature and using them to solve our own problems, we have found a whole new class of computation, a whole new way of using computers. Digital biology will allow us to survive in the modern world. It will guide us through the ever-growing complexity of our global, interactive, fast-paced, modern lives. These new software techniques will provide us with invaluable assistance from their digital universes. They will find us information, detect crime for us, identify faults, and even repair themselves. They will design new products for us, create art, and compose music. They will have originality, creativity, and the ability to think for themselves. How do I know that these things will happen? Because they already have. Using the methods of digital biology, we have achieved all of these feats.

And in the future? Who knows? But it seems almost certain that the first forms of alien life we see will be not through telescopes but through the windows of our computer screens into digital universes. The first person to hold a conversation with an alien intelligence will not be an astronaut, it will be a computer scientist or computational neuroscientist, talking to an evolved digital neural network. The first glimpses of nonhuman cultures and technologies will occur in our research labs, where the digital biology grows more complex day by day.

Perhaps these grandiose visions will be a long time in coming. But the next time you hear your cream-colored box whirring and clicking to itself, just stop and think what type of digital biology might be blooming inside.

WHAT THIS BOOK IS ABOUT

Think I'm kidding? Or exaggerating? Well, read on. The whole point of this book is to explain, for the first time, how biology and computers have become so closely entwined. Chapter by chapter, I describe how the processes of nature work, to the best of current scientific knowledge. I'll include the voices of the biologists who have discovered and investigated aspects of biology. I'll also tell you how computer scientists, designers, engineers, artists, and many other people make use of the same processes with their computers. You'll read how they have used biology to improve our technology, enabling remarkable new advances in all fields. You'll also read how our use of biology within digital universes is expanding our knowledge of life, the universe, and everything. This new breed of scientist, whom I shall call the digital biologist, creates digital universes, genes, evolution, brains, insect swarms, plants, immune systems, and growth. In doing so, digital biologists learn how these processes work.

My aim in this book is to promote understanding. I hope that when you have read it, you will know the reason that so many of us devote our time, energy, and skills to the development of various aspects of digital biology. It is because of the compelling and overwhelming excitement we feel as we uncover some of the fundamental truths of nature.

The book is organized into seven major chapters that explore natural and digital biology: *Universes, Evolution, Brains, Insects, Plants, Immune Systems,* and *Growth.* Each chapter is designed to stand alone with its own distinctive identity. However, if you truly wish to follow the processes of biology that this book describes, you should read each chapter in order. You should then discover, as I have, how all biological processes are aspects of a single, fundamental process, as described in the final chapter: "Answers."

Our first step is the beginning of a theme that I shall continue throughout this book: natural and digital biology follow the same processes, just in different universes. Before you can really understand either, you should understand what a universe is.

2

UNIVERSES

THE UNIVERSE WAS A SWARMING MASS of movement. Like the splashes created by raindrops on a lake, ripple-like things grew quickly into existence. Some had regular shapes; others had highly variable and complex characteristics. As the things traveled, they changed, before slowly fading away. The ripples continuously popped into existence, most expanding in size as they moved. Many resembled little explosions as they grew dramatically in all directions. Others resembled flying worms or forks of lightning. They all flew at similar speeds as they chased their companions into the void. There was clearly a maximum velocity, however, with most approaching the limit but none able to exceed it.

The universe followed slightly different rules from our own. Although there were time and space, it contained no solid matter, no light, no gravity. The only objects that could exist were all constantly coming into being through some unknown agency—perhaps in another universe. And the ripple-like things were traveling *through* each other without any effect. There were exceptions to this rule, however. When two identical ripples traveled in exactly opposite directions and flew into each other, they would both cease to exist. When traveling in the same direction and overlapping, the ripples would merge into larger super-ripples.

Despite the beauty and variety of this universe, it was barren. It contained no life. Nothing born to this cosmos could perceive the seething and never-ending activity. The universe and the ripple-like things within it were destined to an eternity of pointless, mindless existence and destruction.

16

A tragic tale maybe, but this universe is not fictional. Not only does it exist, it is a universe that we have access to. In fact, as you read this, you have sensors dipped into this universe like a periscope, feeding you vital information. I am referring to the universe of *sound*.

As you read this paragraph, activate your periscope: pay a little more attention to the sounds your ears are detecting. If you are in the city, perhaps you can hear the sounds of traffic, a siren, a pneumatic drill, a car horn. Perhaps you hear the sounds of colleagues talking or your children squabbling or crying. Maybe you can hear the music from your CD player while you read this. Perhaps there is just the noise of birds singing or a clock ticking. Whatever you hear, you are experiencing a universe quite different from the one we exist in—a universe comprising only the ripple-like objects that we call sound waves.

This chapter is not about sound; it is about universes. Although it may seem difficult to imagine, the concept of a universe is broader than you might think. There are frames of existence other than the physical universe we experience every day. There are many universes that coexist with or overlap our own. The universe of sound is one, as is the universe of ideas. The digital universe of the computer is another.

We're looking at universes because they help us to understand the full potential of computers. This book is not about the *simulation* of nature or the creation of *virtual* or *artificial* nature. This book is about the use of concepts from nature in a *different universe*—a digital universe. Only by regarding a computer program as a universe in its own right can we take seriously the things that develop and grow within it.

OUR UNIVERSE

How can a computer program be a universe? It sounds a bit unlikely, after all. First let's look more at our own physical universe. It turns out that our universe is a little implausible too.

Using our natural perceptions, we can see much of the universe that affects us. This is no coincidence; evolution has ensured that this is the case. We are aware of solids, liquids, and gases. We can feel the effects of gravity, see light, and feel heat. Nevertheless, there is more to the universe than the things we can see or feel.

Using science, we have broadened our biological perspective of our universe. We know that we live on a (more or less) spherical planet, despite the fact that it looks flat to us. We have worked out that our planet orbits a massive sun. With telescopes we have discovered that there are other planets that also orbit our sun. We know that there are anything from 200 to 500 billion other suns in our galaxy, the Milky Way. In the part of the universe we can observe, there are around 100 billion galaxies. We think that there are a few trillion galaxies in the whole universe. It seems that our universe is *big*.

There's more to the universe than a lot of matter, however. There's also a lot of interesting behavior. Moons orbit planets. Planets orbit stars. Galaxies orbit around themselves and around other galaxies. Light travels very fast, but never faster than a particular speed limit. Matter can emit radiation, and it can be transformed from one substance to another. Electricity can produce magnetism; magnets can produce electricity. Energy can be transformed from light to heat to movement.

Our universe also seems to have a kind of scaffolding to hold everything together. It has spatial dimensions of width, height, and depth, allowing position and size to exist. It also has the fourth dimension of time, allowing past, present, and future to exist. And we know that this scaffolding is a bit bendy. If there is excessive mass (e.g., a star or even a black hole) on one part of the scaffolding, then position, size, and time can become severely distorted.

And as if bendy time weren't enough (although it is normally referred to as Einstein's theory of relativity), it gets worse. Solid is not really solid at all in our universe. We now know that the universe is like an overly ambitious Russian doll. Everything we can touch is made from tiny pieces called molecules. The molecules are made up from even tinier pieces called atoms. The atoms are made from still smaller pieces called electrons, protons, and neutrons. Protons and neutrons are made from even smaller things called quarks. And we think that quarks (and electrons) are points of energy. So the chances are that energy and matter are the same thing.

It seems that our universe is immensely huge and made entirely from energy that seems to interact with itself according to certain rules. If only it were that simple. The trouble is that some things in our universe behave in very peculiar ways. When you perform even quite simple experiments with light, some odd interference patterns emerge. The interference looks exactly like the kind of thing one would expect if other light sources were present, *but there is nothing else there*. We see such mysterious interference even when we do experiments with one photon at a time (light is made from photons). These and other effects have led to the development of quantum theory: that the interference of light looks as though it is caused by the photons in other universes influencing the photons in our own. If we work out the most extreme repercussions of this theory, then we must conclude that there are other universes, parallel to our own. There are parallel versions of me writing this and parallel versions of you reading this. And it seems that there may be an infinite number of these parallel universes out there.

As you may have noticed, this area of science strongly resembles science fiction. There are many other things we know about the universe that are equally hard to accept—the fact that time does not flow, for example. Just as we do not flow through space, our impression of an ever-changing now that moves through time is illusionary. We exist in space-time—a four-dimensional universe comprising all locations and all times. Some theorists assert that every concept of now is exactly equivalent to a parallel universe. But let's not pursue this any further. I have just one more mental punch to throw at you. This one is about the origin of our universe.

One thing we did notice fairly early on was that all the galaxies in our universe appear to be flying away from each other at a considerable rate. When we traced the paths back, it became clear that everything in the universe seemed to originate from a single point in space. This led to the big bang theory, which proposes that a large explosion resulted in the raw components from which everything we see today emerged. Not only did the explosion generate the energy and matter required for the universe, but we think it also laid down the space-time scaffolding. Before the bang, there was no space or time. We do not know for sure exactly what exploded, but it was probably a hyperdense point of energy, the size of a quark. We're also not sure where it came from. Some theories state that there may have been a period of chaos during which certain quantum laws still applied, resulting in the point of energy popping into existence. Certainly for the foreseeable future, the big questions of how and why our universe came into being will remain unanswered.

This is our physical universe—or at least as much of it as we can examine, calculate, and guess at. It is a very different place from the one we actually observe in our day-to-day lives. Trapped in our physical universe as we are, we cannot see far enough to determine the true nature of our home.

WHAT IS A UNIVERSE?

LAWS

Now that you have a sense of what a universe can look like, it is time to generalize. Before you can believe that there are other kinds of universes out there and, more important, that some of them exist in our computers, you need to know what a universe actually is.

Some regard a universe as a box or container that holds everything. Others think of a universe as everything that is held in the box. In fact, a universe is both. Let's build a small, imaginary universe to illustrate this.

First, we need to define a few laws. If we're going to have things as part of our universe, they need somewhere to exist. Let's create two spatial dimensions and a dimension of time. This space-time scaffolding is our "container" in which we can put things. Should anything exist, it is now possible for it to have a spatial location and movement. Let's also impose a shape to space-time. We'll make it into a torus: anything that goes too far to the right finds itself on the left, anything moving too far to the bottom will end up at the top, and anything existing toward the end of the universe (in terms of time) will end up at the beginning.

It's time to put things in our universe. We can make up the idea of energy with another law and add lots of it—perhaps lots of little energy points. If we add the right laws, we can make those points clump together, transform one another, induce movement, and have other interesting behaviors. If we're really clever, we can find the right laws to ensure that emergent phenomena occur—complexity arising from our simple components. Perhaps our energy will begin to behave like atoms. If we add a law that makes space-time distort in the presence of high levels of energy, maybe we can induce effects resembling gravity.

We now have a new universe. Its "container" is the space-time scaffolding. Its "contents" are the energy points. And as we made this universe, you may have noticed a running theme. We needed *laws* to make our universe. To create space-time, we had to invent an appropriate law. To allow energy to exist and do something, we needed other laws. A universe can exist only with laws.*

What do we mean by a law? What is a law? In our society, an example of a law might be "If you steal, you will be punished by imprisonment of up to ten years." We understand this to mean "If you are caught and convicted for theft, you may receive a sentence of up to ten years in prison, although you will probably be released early for good behavior." Laws in our society tend to be a little unreliable and inconsistent in their implementation. In a universe, laws of physics are never unreliable or inconsistent (if they appear to be, then we do not understand the law properly). An example of a law in our universe is "Objects distort space and time in proportion to their mass." Assuming this is correct (and to the best of our current knowledge, we think it is), then this is a fundamental property of the universe. The law is unbreakable: it cannot be adjusted or interpreted by clever lawyers. It is literally a part of the universe.

This is an important point that helps to define what a universe is. A universe cannot exist without laws. Laws define the fundamental properties of the universe, including its existence. Universes are made from laws.

RULES

There is another word we could be using instead of *law*. The word is *rule*. The laws of physics that define our universe can be expressed as rules. And we know a thing or two about rules. A rule takes the form:

If something, then outcome.

The rule tells us that every outcome depends on something. Without that something, the outcome would never happen. With it, the outcome will always happen. Every outcome is caused by a something. This is a

---

*Many scientists think that there may be a single fundamental law of everything that produces all the effects in our universe that we currently attribute to many different laws. Nevertheless, whether the universe has one or many laws, the principle is the same.

simplification, of course. Most real rules define the outcome to vary according to the something, and some rules may even be recursive, with the something being some aspect of the outcome. But this does not matter. The point is that all rules specify that outcomes must depend on something. If they don't, they are not rules. So if our universe is defined by rules, which it is, it follows that each of those rules operates on something to cause an outcome. If I push something, then it will move (or I will). If there is a large mass, it will distort space and time. If energy exists, it can never be lost, only transformed. There are some of the rules of our universe.

We can also learn more about the kinds of rules that define universes by thinking about how they are created. Let's look at the creation of our universe. If we try to find an ultimate cause for our universe, clearly we need a rule that says, "If there is nothing at all—no space, no time, no energy, no other universes—then create a super-dense point of energy and a set of other rules to go with it." Unfortunately, this does not make sense. If there is nothing at all, then where did that rule come from? What created it? Another rule? And what created that? Our universe does not make any sense.

To help understand this, let me show you a universe that does make sense. There's nothing up my sleeve . . . abracadabra! Here it is. It is a null universe.

The laws of the null universe are simple—there are none. It has no space-time scaffolding, no energy, nothing, with no behavior. It is a universe without a box and without any contents. It is nothing.

No matter how uninteresting the null universe is to us physical universe dwellers, it does at least make sense. The null universe does explain its own existence. There is nothing there because there are no rules to allow the creation of anything. There is no behavior because there are no rules to give any behavior. There are no rules because there was nothing there to create, derive, or produce the effect of rules. Unlike our own universe, the null universe can explain its own (non)existence *in terms of itself.*

We simply cannot explain the existence of our own universe without some external *something* triggering it into being. And we do need to explain it, for, unlike the null universe, there is a lot to explain. As soon as there is anything—a rule or energy or behavior or matter—we must ask, *why?* Even if the rule is the only conceivable rule, the only possible rule that can define a universe, the only way to know why the rule is there at all, and indeed, what is making the rule actually do anything, is to define the universe with respect to something else. In the same way that you cannot tell if a new color "works" without first having a color scheme, a new universe cannot "work" without first having an existing "scheme."

It seems that this feature is common to all universes except for the null universe. All other universes have laws. In some cases, they define space-time and energy and distortion; in other cases, they define other things. But we will always need an ultimate law or rule to explain what caused those universes, and that rule must always explain the causation in terms of something external to the universe it is explaining. It is not good enough to answer the question "Why does something exist?" with "Because it has always existed." The answer does not explain the *cause*. It does not explain why the null universe is not here and we are.

Here are some examples of answers that do explain the cause of our universe:

- A chaotic universe prior to ours had a rule that resulted in the creation of our universe.
- A parallel universe "donated" a piece of itself that became our universe.
- God decreed that we should exist.
- A dragon sneezed, and we all came into being.

All of these answers, however believable, rely on an external universe of some kind. Every other answer will also do so, or it will not explain why we are here and the null universe is not.

So it may be that except for my null universe skulking in the corner, it is not possible to explain or even define a universe in terms of itself. Universes are simply not self-contained. They rely on other universes to cause them and make them what they are.

This completes our search for a definition of a universe. From this exploration, we know that a universe is made from a set of rules that define outcomes. We also know that at least one of the rules must be defined in terms of another universe. This two-part definition acts as a key that opens the door to other universes. Now that we know what a universe really is, we can explore some of the large number of universes that coexist with us.

OTHER UNIVERSES

Many other universes exist and overlap with our own. Some we are aware of; some we are not. Some we help create; others we do not. But how do we recognize another universe?

We know that a universe consists of rules, one or more of which must be defined with respect to a different universe. With this definition, we can start to think about other universes. If a set of rules defines our universe, other universes will be defined by other sets of rules. So if we can identify an environment defined by different rules from those that define our universe, *it* may be a different universe. We began this chapter immersed in the universe of sound. Let's briefly return to it and check that it does deserve to be called a universe.

The universe of sound is similar to our own universe. It shares rules that define a very similar space-time scaffolding. It also has some different rules. In our universe, the maximum velocity of everything is the speed of light. In the universe of sound, the maximum velocity is much slower—it is the speed of sound. In our universe, gravity plays a highly significant role. In the universe of sound, there is no clear concept of gravity. In our universe, most things cannot pass through the same space without colliding with each other. In the universe of sound, all things (sound waves) can normally pass through each other without any effect.

Clearly the universe of sound has some different rules from our laws of physics. It is also clear that the rules for the universe of sound are defined in terms of our laws of physics. To an imaginary dweller of the universe of sound, the ripple-like objects spontaneously appear from nowhere. To us, we can see that sounds are generated from vibrations of our molecules in our universe. So the rules defining the universe of sound are defined with respect to another universe—our own. (This is surprisingly similar to those mysterious interference patterns we observe with light. To us, they appear to be spontaneous and without explanation. If the theories are right, then to the dwellers of other universes parallel to our own, our interference patterns are caused by their photons interacting with ours. So our laws of physics may also be defined with respect to other universes in this way.)

Because the universe of sound has different rules from ours and because some of its rules are defined with respect to an external universe, it does qualify for the designation of universe.

Not all universes share quite so many rules with our own. The universe of ideas, or memes, is quite different from ours. The memetic universe, as I shall call it, has a very different scaffolding from ours. Objects (memes) in the memetic universe have no position or size; there are no spatial dimensions. There is a dimension of time, however, so memes can exist at certain times but not at certain locations. Because of this, there can be no movement, no speed, no acceleration, no collisions. But there are some exotic rules that define when memes are created. It seems that as we travel forward along the time axis of this universe, we see more and more memes in existence. Perhaps one rule of this universe is that the number of memes increases as time passes. Other rules also seem to affect this universe. New memes are always very similar to existing memes except for slight variations. Some memes seem to persist for long periods; others come and go quickly. At times, a specific property of a meme will cause many new memes to be created soon after, like a population explosion.

Because the memetic universe is so different from ours, it is very hard to imagine that it might really exist. If it does, it will be because the theories of memetics are somehow shown to be true. This viewpoint suggests that ideas, concepts, even catchy tunes exist like parasites in our minds. We call them memes, and the ones that we favor become more numerous, while the memes we dislike slowly fade away. This is an evolutionary process—the evolution of memes within our combined brains, books, and other media. Memetics requires us in our physical world to copy, alter, and select memes. But the memes themselves exist in the memetic universe (sometimes called the *meme pool*), with its unfamiliar rules that are defined in terms of our own universe.

Enough of memes and sound waves—there are many other universes out there. The universe of mathematics, for example, where strange symbols are constantly transformed by rules.* The universe of finance, where mysterious rules determine the flow and transformation of different monies. Biologists such as Richard Dawkins explain evolution in terms of the universe of genes (although they may refer to this universe as the gene pool), where objects are genes that change according to evolutionary rules. Each one of our games is a tiny universe in its own right, defined by rules that relate to our own universe. Every chess game involves the creation of a new "chess universe," with the players implementing the "laws of physics" for that universe.

Many universes coexist with our own—some large, some small; some complex, some simple. However, there is one type of universe that is unlike any other. Similar to the memetic universe, it is created by us. But unlike the other nonphysical universes we have explored so far, this universe can support complex processes such as growth, thought, and life. I shall call this exciting new cosmos a *digital universe*.

DIGITAL UNIVERSE

Everything is dark. Suddenly, in a flash, the universe is full of activity. We see hundreds of little red boxes, scattered randomly around the otherwise dark landscape. We see a small gray circular object in the center. And all around this object are little yellow things, scurrying about at great speed.

As we watch, the antlike things explore their universe. Some seem to be trying to escape, but they are prevented from doing so by invisible boundaries around the environment. Others meander at frantic rates through the landscape. Whenever one comes across a red box, it seems to pick it up and carry it around. Soon many of the yellow "ants" are carrying red boxes as they rush around in all directions.

Once in a while, a box is put down. When a box has been dropped, it is never picked up again. The antlike things seem to "glue" the boxes to the ground and the color of such boxes changes from red to blue.

---

*Since I wrote this, I discovered Ian Stewart's delightful book *Nature's Numbers,* in which he describes a mathematician's view of the "mathematical universe," as he calls it.

Soon a pattern of blue boxes becomes evident among the red. As more and more boxes are carried and glued down by the "ants," a distinct circular wall of blue boxes begins to form around the little gray object in the middle. Despite the apparent random scurrying of the yellow "ants," this wall soon becomes complete as the final red blocks are carried and glued into the last few gaps in the wall. Their task accomplished, all movement ceases. Everything in the universe disappears.

This high-speed universe with its short lifetime has been created and destroyed many times. Similar yellow ants have constructed similar walls many times previously. Indeed, as I write this, the universe has just come into existence again, with its red blocks and scurrying yellow ants going about their mindless activity.

As I'm sure you have guessed, this universe exists in my computer. It is a digital universe. I can never touch a yellow "ant" or pick up one of the red boxes. I will never be able to walk around the smart blue wall and admire its construction. These objects exist in another universe from me. But the fact that they do exist is undeniable.

The digital universe described above has three dimensions: two spatial dimensions and a dimension of time. This simple space-time scaffolding supports a number of objects, where each object has a spatial position at a specific time. We know that the universe is not infinite. We can observe its creation and destruction, so we can see that it exists for only a finite amount of time. We can also see that there seem to be boundaries around the space of the universe, so space is also finite.

Although every object in the digital universe has a position in space, only some can move or be moved. The little gray circular object in the center never moves. The antlike things always move. Red boxes can be moved by the ants. Blue boxes are immovable. And if we could perform some experiments on the objects in the universe, we would find out even more. For example, if we could move the little gray circular object, we would discover that the blue wall would be built in a different place with the gray object still in the center. If we could remove the gray object altogether, we would discover that no wall would ever be built; the ants would never drop their red boxes. With sufficiently ingenious experiments, we might discover why the wall is built at all. We would find out that the little gray object is continuously shouting messages at the yellow ants. The invisible messages inform each ant how far away it is from the gray object. We would also discover that when the ant is carrying a box and happens to be a certain distance away from the gray object, it is instinctively compelled to glue that box to the ground. The ants wander randomly all around the gray object, and all of the ants glue down their boxes when they are the same distance from the object. But they cannot walk through boxes or other objects. These simple rules mean that boxes are always glued in a space on a circle around the gray object. Because there are many ants with many boxes, the behavior results in a circular wall.

These are some of the rules that define this universe. I know that they are correct, because I created those rules. I thought of them, and I wrote them down. I couldn't use English, though, or mathematics. I used a programming language. Why? Because the collection of rules that define this little digital universe is known as a computer program.

MACHINES TO CREATE UNIVERSES

Computers are not like our other machines. Traditionally, each machine has only one behavior. Clocks are often intricately designed, but they only do one thing. They only tell time. The components all do what they were designed to do, but a clock can never vacuum your carpet, just as a vacuum cleaner will never tell you the time. The engine in your car will make the wheels move, but it will never be able to record your voice on a tape. Each of our machines is designed to do something, and that is all that they can ever do. But computers are different.

Computers are machines that can behave in any way you like. Indeed, this is the function of a computer. We don't use computers in order to shuffle electrons around the place, although that is all that they actually do. We use them to behave in ways that are beneficial to us. That is why computers always need software. Without their programs, computers have no behavior. A computer program is the set of instructions that give the computer a behavior.

Every computer program is essentially a collection of rules that defines what can exist and provides a set of behaviors. In other words, each computer program defines a new digital universe. When the computer runs a program and those behavior-defining rules are executed, the computer becomes a universe generator. The program defines the laws of the digital universe, and the computer causes that universe to come into being.

Before we can really explore the nature of digital universes, we need to understand computers a little more. What kind of machine can follow instructions and potentially do different things every time it is used? What kind of machine can generate universes? How do these machines work?

HARDWARE

There are two aspects to every computer: the hardware and the software. Computer hardware is the stuff you can touch, smell, and thump. Computer software is the stuff you feed to the computer to specify behaviors (and define digital universes).

To help me explain how computer hardware works, let me introduce you to four friends of mine: Mary, Al, Reg, and Ian. Mary has a wonderful filing system. Ask her for any information, and she'll provide it after a quick dig around. Al is a bit obsessed with his codebook. Although it may contain only a limited repertoire of codes, he always works out what everything means and gives his friends the results. Reg has an amazing short-term memory: he can repeat small amounts of information in a split second. Finally, Ian is a bossy fellow. He's forever telling Mary to remember things and waiting for information.

My four friends always play the same roles in a conversation. Here's an example of the kind of thing they say to each other. Because Ian is so bossy, he always begins first:

Ian: Mary, please keep the following seven numbers for me in your filing cabinet: 100, 010, 001, 101, 110, 110, 000.

Mary: Okay, Ian.

Ian: Al, why don't you start doing something?

Al: Okay, Ian. Mary, can you tell me the first number that Ian gave you?

Mary: It's 100, Al.

Al: Thanks, Mary. Well, that's interesting. According to my

codebook, that means *add the next two numbers together.*
Mary, do you want to help me out here?

Mary: The next two numbers are 010 and 001, Al.

Al: Right. Well, according to my codebook, the sum of 010
and 001 is 011. Reg, would you remember the number
011 for me?

Reg: Got it, 011, no problem.

Al: Thanks, Reg. Mary, would you tell me the next number,
please?

Mary: It's 101, Al.

Al: Right. According to my codebook, that means *store Reg's
number in the place given by the next number.* Mary, I need
the next number, please. Reg, could you tell me your
number?

Mary: It's 110, Al.

Reg: My number is 011, Al.

Al: Thanks, you two. Mary, could you store 011 in the file 110,
please.

Mary: Okay, Al.

Al: Thanks, Mary. Could you give me the next number, please?

Mary: It's 110 again, Al.

Al: Hmm . . . According to my codebook, that means *tell Ian
Reg's number.* Reg, could you remind me what your number was again?

Reg: It was 011, Al.

Al: Thanks. I have a number for you, Ian. It's 011.

Ian: Is it really? Thank you, Al.

Al: No problem. Mary, could I have the next number, please?

Mary: It's 000, Al.

Al: Well, according to my codebook, that means *stop.*

At this point, all conversation usually ceases, until Ian starts talking again.*

---

*To avoid excessive tedium, I've cut out a bit of the conversation. In reality, Al
will usually ask Reg for information, who will ask Mary. Al doesn't like to
speak to Mary directly all that often.

Mary, Al, Reg, and Ian are not people, of course. They are the basic electronic circuits that make up the hardware of your computer. Mary is the *memory*. Information in the form of binary 1s and 0s is stored by high or low voltages. These are held constant using tiny electronic batteries called capacitors and changed using electronic switches called transistors. This kind of memory is often called RAM, or random access memory. The cryptic acronym just means that we can access any snippet of information instantly, without needing to spool through vast amounts of other data, as we would have to with magnetic or paper tape.

Al is short for *arithmetic and logic unit* (ALU). This electronic device produces various electrical outputs (results) that depend on electrical inputs. The behavior of the ALU is fixed and unchanging. The electrical output for every instruction-data combination is hard-wired, in much the same way that pressing the *s* key on a typewriter causes the letter *s* to be stamped on paper.

Reg is short for *registers*—temporary storage areas that work in the same way as the memory (but access is considerably faster) designed to hold the results from the ALU and the memory.

Finally, Ian is the I/O circuitry, which provides the interface from input devices such as keyboards and the interface to output devices such as monitors and printers.

The "conversation" provided above was an example of a *computation*. The seven numbers provided by the I/O circuits (my friend Ian) and stored in the memory (Mary) made up the program: a combination of instructions and data. The fetch-execute cycle then began. The ALU (or Al to his pals) requested each number in turn, decoded the instruction, and executed it (sometimes by requesting numbers from the memory and the registers). The results of execution could be anything from storing numbers in the registers or memory to outputting a result. Once executed, the ALU requested the next number from the memory, decoded it, executed it, and so on. The fetch-execute cycle ends only when an instruction tells the ALU to stop.

It should be clear that our friend Al plays a fundamental role in every computation. The ALU is the circuit that carries out all instructions. Its electronics transform every combination of instruction and data into the appropriate output. Essentially, the ALU is following rules that are embedded into the design of the circuit. If the instruction is *add* and the data are 010 and 001, then the electronics will manipulate those electrical signals and generate the output of 011. Alternatively, if the instruction is *add* and the data are 011 and 001, the output of 100 will be produced by the ALU. The execution of every instruction follows a rule (or set of rules) that has been integrated into the circuitry of the ALU.

This is the underlying machinery of a digital universe. The electronic cogs of the computer turn predictably. Every action is the result of electronic circuits directing the flow of electricity. The intricate dances of electrons tracing their way through the myriad circuits that make up our computers are all caused by electronic rules.

SOFTWARE

Computers are machines that perform instructions according to rules. Most computers have between 50 and 200 different instructions in their instruction sets. The instructions usually operate on between zero and four binary numbers at a time, where each number can be anything from 8 to 128 bits long, depending on the word size of the computer. Clearly, the number of different data instruction combinations is an astonishingly large number. (I was certainly astonished when I worked it out.) But however huge this number is, it is not infinite. And yet a computer has an infinite number of different behaviors.

The reason that the behavior of a computer is infinitely variable is that we can choose which instructions the computer should execute, and we can choose the order in which the instructions are carried out. In this respect, instructions are like musical notes. Just as a finite number of different notes can be used to produce an infinite number of different melodies, a finite number of different instructions can be used to generate an infinite number of behaviors. And we tell the computer which instructions it should execute by giving it a *program* to follow.

We have already observed an example of a simple computer program. The conversation among Mary, Al, Reg, and Ian began with Ian providing seven binary numbers. These were examined in turn by Al, who decoded and executed them. The numbers formed a program.

These days, we do not write programs in machine code or specify which of the low-level instructions should be executed in our programs.

Instead most of us write programs in high-level languages. These programming languages (such as Pascal, C, C++, Lisp, and Java) are designed to resemble our own language more closely. Using the programming language of our choice, we write computer code in "sentences" so that we can read and understand it ourselves. We then use *compilers* (other programs) to translate our programs into machine code, which the computer can then execute by following the rules embedded in the circuitry of the ALU.

So we define our digital universes using high-level languages. Although every command is executed by the low-level rules within the ALU, more interesting higher-level rules define the digital universes that we shall be exploring in this book. These rules are defined by the commands of the high-level language. Every IF statement, every FOR loop, and every variable declaration helps form the rules that define the current digital universe.

But this does not mean that every programming language defines a different set of digital universes. In fact, it does not matter which high-level language we use. The same high-level rules can be expressed in nearly all of them. To see this, let's create a very simple digital universe using the high-level language C:

```
void main ()
{    int position;
     for (position = 0; position < 80; position++)
     { gotoxy (position, 10);
       printf ("digital nature");
       getch ();
     }
}
```

Here's the same description of a digital universe, this time created by the high-level language Modula 2:

```
MODULE example
VAR position : INTEGER;
BEGIN
     FOR position := 0 TO 80 DO
```

```
GotoXY (position, 10);
WrStr (` digital nature');
REPEAT UNTIL KeyPressed ();
END;
END example.
```

When we compile either piece of code and execute it on the computer, we see a black screen with the white text "digital nature" halfway down on the left. Every time we press a key, the text moves to the right by one character. We can control the speed of the text by how quickly we press a key. When the text has disappeared from the right of the screen, the program ends.

Both programs do exactly the same thing: both define the same digital universe. To understand this, let's look at the laws of physics for this universe. Here are the high-level rules that define this very simple digital universe:

When something creates the universe, two spatial dimensions capable of supporting objects called *characters* will be created.

If the universe has just been created, then "digital nature" will be written halfway down on the left of the screen.

Every time the user presses a key, the text will move one character to the right.

Once the text has disappeared from the right of the screen, the universe will be destroyed.

The digital universe in the example has only four high-level rules. Both programs follow this set of rules, so both programs define the same digital universe. It is quite normal to describe the working of a program as a set of high-level rules, written in English. Such a description is called an *algorithm*. From now on in the book, we will look only at algorithms; we will not be concerned with which high-level language is used to write down that algorithm.

But can we legitimately call the execution of this algorithm a universe? Recall our definition: a universe is made from a set of rules that define outcomes. At least one of the rules must be defined in terms of another universe. It should be apparent that the algorithm is a set of rules that define outcomes. It should also be clear that any rule that refers to input (a key being pressed) or output (writing text on a screen) must refer to something outside of the digital universe. (In addition, the very creation of the universe requires the existence of another universe.) From our definition, we can see that a computer program does fulfill the criteria. A computer program defines a digital universe. A computer executing that program creates that digital universe.

This is the essence of software, computer programs, and digital universes. Fundamentally, they are all a collection of rules that define the behavior of the computer. These rules are written down using a high-level language, complied into a much greater number of low-level instructions, and executed by the rules embedded in the electronic circuits of the computer.

VIRTUAL MACHINES ARE REAL UNIVERSES

Universes are made from laws or rules, and a computer program is a set of rules. When a program is run, when the rules are followed by the computer, a digital universe is created. In computer science, this is not a new idea. The concept of a digital universe has been known and understood for many years. It has simply had a different name. Traditionally, digital universes have been called *virtual machines*.

To understand what a virtual machine is, let's return to the birth of computer science. We need to see the Universal Turing Machine (UTM), a piece of conceptual machinery that mathematician Alan Turing developed in 1936.* Although it remained an unbuilt abstraction, it was the seed from which all of today's computers grew. With the help of fellow mathematician and Turing's Ph.D. supervisor, John von Neumann, an extended version of the UTM, known as the *von Neumann architecture,* became the blueprint for all computers today. Indeed, my four friends, Mary, Al, Reg, and Ian, illustrated the action of a von Neumann machine in the example provided earlier.

---

*Turing actually created the UTM in order to show that there is no algorithm that can, in principle, decide all mathematical questions. We won't worry about that here, though.

The UTM is showing its age a little now. Its operation revolved around paper tape and a read/write mechanism. On the paper tape was a long sequence of 1s and 0s. The UTM could spool the tape in either direction, read the current digit, or write a 1 or 0. The paper tape was used as the memory for this conceptual machine. It could store data and simple instructions. These could be read and decoded, and they would tell the UTM to perform basic operations such as *read, erase, write, move left one space,* and so on.

Although this computer seems a little primitive compared to modern technology, as a mathematical model of computation, it was more than sufficient. Indeed, Turing proved that given the right set of instructions, the UTM could duplicate the behavior of any other computer. This remains true to this day. The UTM can perform the same instructions as our most powerful computers simply by shuffling its 1s and 0s about in the appropriate ways. (It won't be as quick or as efficient as the computer it is emulating, but it will have the same behavior.) The reason this is still important is that the underlying architecture of all of our computers is based on the UTM. So every one of our computers is capable of emulating the behavior of every other computer. My computer can behave in exactly the same way as computers designed twenty years ago. (Indeed, I have software emulators that allow my computer to run the software of twenty-year old computers.) In the same way, a twenty-year-old computer can emulate my computer (although it will be much slower and may require more memory).

When a computer is asked to run such emulation software, it is transformed. It can no longer run the software designed for it. The action of the keys, the output to the screen, the operation of the mouse—everything has changed. The computer can now run only software designed for the computer it is emulating. Because the emulation software alters the behavior of the computer in this way, we call it a *virtual machine.*

In general, a virtual machine is a piece of software that defines an environment. That environment may duplicate the behavior of another physical computer or may be a distinct environment in its own right—an environment that can be created by different models of computers and provide a consistent interface for other pieces of software. And with a little stretch of the terminology, we can regard every piece of software as a type of virtual machine. Your operating system defines a clearly identifiable environment, a familiar look and feel, with consistent behaviors in response to your input. Your word processor defines another environment, as does your spreadsheet or computer-aided design package. Our computers are capable of supporting many such software-generated environments to help us work. These can all be regarded as virtual machines.

So the experience of a virtual machine is very familiar to everyone who has ever used a computer. And clearly, virtual machines and digital universes are the same. Virtual machine is one name for a piece of software. Digital universe is another. Every time you manipulate files with a Windows-based operating system or edit words in a word processor, you are interacting with objects in a digital universe.

Digital universes have existed since the invention of the Universal Turing Machine. Until now, they have simply gone unrecognized.

MONOPOLY MONEY?

Universes are defined by rules. Computer programs are defined by rules. At least some of the rules are defined with respect to other universes. The rules of both specify environments. The similarities are too great to ignore. Digital universes exist.

It is at this point I take out my banner with *Equal rights for all universes* written boldly across it. Digital universes get a very rough deal, you see. Words such as *artificial, simulated,* and *virtual* are always used as superfluous companions to anything resembling our own universe that we see in our computers. We hear endlessly of virtual realities, artificial life, simulated environments. Why?

Why should something in the digital universe of a computer be considered less real than a physical object in our universe? Why is virtual reality *virtual*? Why is artificial life *artificial*? Isn't that a little unfair?

As I will explain in the next chapter, when I run an evolutionary algorithm on my computer, it does not simulate evolution. My computer allows real evolution to occur in a different medium. And when my computer creates an environment with mountains and valleys and digital organisms with behaviors, from the perspective of the inhabitants, the environment is not virtual; it is real. If we ever manage to evolve a neural network capable of consciousness in a digital universe, will it not deserve to be called alive rather than artificial?

Regarding the behavior of a computer as somehow unreal is a mistake made by far too many. But when we think about what the computer does and how it does it, such a view is untenable. When my computer multiplies two numbers together, does it pretend to do the sum? Does it simulate mathematics? Of course not. The ALU follows the rules of addition embedded in its circuitry and calculates the correct answer. In contrast, when *I* multiply two numbers together, I do not always calculate the result. I will often rely on my memory of the correct number. I am simulating multiplication by using memory rather than the rules of mathematics. So for mathematics, the behavior of my computer is *more real* than my own behavior. I am an artificial calculator.

I am currently reading text on my computer screen. The text is made from lots of little dots, placed in the right patterns to make the shape of letters. Newspapers and magazines use the same method. Is the text generated by my computer less real than the text in our newspapers? Is it less real than the illegible scrawl of my handwriting? Of course not.

Banks use computers to store how much money we have in our accounts. The flow and ebb of money between accounts happen by transferring numbers between computers. Does this mean that when we pay money into a bank account it becomes a *simulation* of money? Does it turn into Monopoly money? I hope not!

Digital universes are not simulations. They are not fakes or metaphors. Even their inhabitants are surprisingly similar to us. We saw at the beginning of this chapter that every object in our universe can be broken down into smaller and smaller pieces, until we found that the smallest bits were points of energy. What are digital objects made from? They are made from binary 1s and 0s, held as patterns of electricity: the flow of electrons. What is an electron? A point of energy. So physical objects and digital objects are both made from energy. *We are even made out of the same stuff!*

So I think it is time to think a little harder about the capabilities of computers. These machines are not automated slide rules. They are not clever televisions, glorified drawing boards, or overcomplicated typewriters. They are not even electronic brains. Our computers are much more than this. They are universe creators.

SUMMARY

This chapter has set the scene for the rest of the book. Throughout it, I've tried to encourage you to take a broader view of the computer. Instead of being a machine that blindly performs a single predefined action, the computer is a universe creator.

Of course, our computers are capable only of creating simple universes compared to the one in which we exist. But computers do have one advantage over our universe. With computers we can define the laws of physics. We can write software that defines the fundamental movement of bits inside the digital universe. And if we make our digital universes follow similar rules to those we observe in our universe, we see the emergence of processes, intricately structured objects, and complex behaviors that resemble those we see around us. Digital biology blooms.

The next chapter describes one of the most fundamental processes in nature: *evolution*. We'll see how and why evolution happens in nature, and explore how we use evolution when it occurs in our digital universes.

3

EVOLUTION

THERE IS A CREATURE sitting on the bookshelf behind me as I write this. When I peer closely at it, I can clearly see the facets of its compound eyes, each a tiny lens that seems to be looking back at me. The two eyes are shaped like miniature segments of an orange, protruding from the flat, crescent-shaped head and giving a 360-degree view of the world. Its body, the length of my thumb, is covered with beautifully designed armor plating, segmented and overlapping to provide flexibility. If the underside of this creature were visible, I would be able to see many rows of delicate legs.

Sadly, the creature is not in its natural environment. Its home was in shallow water, where it used its legs to stir up tiny particles of food and pass them to its small mouth. When threatened, it would burrow into sediment and hide from predatory fish. Today, it sits on my bookshelf, currently with its back arched and its head facing in my direction. But its position is not very surprising, for it has remained fixed and unmoving in that pose since its death, 390 million years ago.

Here I must come clean, for this creature is a trilobite; to be more precise, it is a fossilized trilobite. All movement has ceased forever, all life has departed, all living tissue long since substituted by rock. And yet even from such ancient remains as this, we can deduce so much of life. Indeed, we can deduce the method by which all life on this planet was created.

Fossils are nature's photographs, revealing forgotten creatures living and dying in worlds unimaginably distant and unfamiliar to ours. Each world is Earth at a different point in time, illuminated by the flashbulb of discovery of a fossilized creature or plant. We have thousands of these snapshots of past life, each one carefully studied and catalogued. These separate "photographs" can be placed one after the other, ordered according to the time at which fossilization occurred. Together they form a conceptual movie of the development of life on our planet. When played (and for this I'm afraid we must rely on our imagination), we see something astonishing. Life *changes*. At the start of our film, there is far less life and far simpler life. As we watch, the forms of creatures alter, their shapes morphing into different contours. One type of shell-fish becomes six types; some persist; others disappear from view.

Enough of the preview, let's dim the lights and watch this imaginary "fossil film." The first flickering pictures begin around 3.9 billion years ago (that's a staggering 3,900,000,000 years in our past). We see a world with an ammonia- and methane-rich atmosphere before the Earth's crust has cooled enough to allow rocks and continental plates to form. And yet there is life. We see microfossils of blue-green algae. (Fossils resembling these in Martian rocks recently caused a debate about life on Mars.) As we watch, we see more and more bacteria appear, and other forms of primitive life such as archaea, which still live in hot springs and rift vents at the bottom of oceans. The herbivorous eukaryotes also appear and slowly change. Primitive life soon fills the early oceans, the atmosphere of Earth becoming increasingly oxygen rich over the next few billion years. As the first stable continents appear and begin to form the single super-continent now known as Rodinia, a breakthrough occurs: multicellular organisms grow before our eyes. Simple plants, wormlike animals, and fungi develop from eukaryotic cells. These slowly change and diversify into algae, corals, molluscs, starfish, sponges, and the first trilobites. We see ever more complex creatures growing and swim-ming, eating and being eaten in the oceans. The first fish develop, plants begin to colonize land, and spiders and centipedes follow. (If we press the pause button at just the right moment, at around the time when the first trees began to rise above the ferns on the land, when the first insects began flying, we would see my trilobite living its short life. If it had managed to reproduce before it died, then perhaps its descendants joined the countless number that developed into the many different trilobite species that survived for another 150 million years.) Our film plays again and shows the Earth filled with ever-changing life (as well as some mysterious mass extinctions). Continents are torn apart and hurled back together by the movement of tectonic plates. Glaciation causes sea levels to rise and fall, compressing and transforming vast forests into coal. Life continues to survive and flourish. As the time becomes closer to our own, the picture becomes less discontinuous and disjointed. Another supercontinent is formed as lands collide. On Pangea, as we refer to it today, a new type of plant slowly supersedes the forests of fernlike trees. It uses the revolutionary method of seeds to distribute its offspring. Soon forests of early conifers grow. We see the development of reptiles, growing ever larger. It is the time of dinosaurs. Small and great, constantly changing reptilian beasts dominate the land. The stegosaurus grows and disappears; the tyrannosaurus develops and also vanishes. Around them, the entire ecosystem of Earth thrives and changes. Dinosaurs rapidly begin to be replaced by mammals. Flowering plants, insects, fish, and small flying dinosaurs, which we now call birds, diversify. As Pangea is ripped apart to form the continents we are so familiar with today, large mammals develop and disappear. Just before the film ends, we see a hairless bipedal mammal morph into existence.

The lights come up. Admittedly, our imaginary film was jerky and very incomplete in places, but the conclusion is inescapable. The living creatures around us today were not generated and deposited on Earth in one creative burst of activity by some deity. Our planet has been overflowing with continuously *evolving* life for thousands of millions of years.

For a minority of people, this view of the origin of life is deeply disturbing. Indeed, they feel that it threatens their religious beliefs and therefore must be dismissed as an unproven theory. The evidence for evolution is, however, quite overwhelming—not just in our fossil records but, as Darwin discovered on the Galapagos Islands, in the different living species that exist today. Evolution is undeniable. With the discovery of DNA, the mechanisms behind evolution are also known. We have vastly more evidence for the existence of evolution than we do for the claim that smoking causes cancer, and it is a brave individual who tries to dispute that theory.

In the words of Steve Jones, the well-known professor of genetics at University College London, "Biology equals evolution. If you read Darwin's *Origin of Species* (or if you read my revised version) . . . what's so fantastic about the *Origin* is that Darwin went out and picked up apparently unrelated pieces of information, like birds on islands, say, or domestic dogs, or fossils from South America, or plants, pigeons. Just all kinds of things that before, nobody really saw as bolted together into a coherent whole. He made them into a single paradigm which is linked by evolution. Basically now, whatever you're doing in biology, you're really studying evolution. It's very noticeable that lots of biologists do that without even noticing it."

But the point of this chapter is not to try to convince anyone of the existence of evolution. Evolution not only exists, but we use it every day on our computers. The point I'm making here is about power. Nature provides us with an unrivaled view of evolution in action. And natural evolution is an immensely powerful process. It operates on geological timescales and yet can also cause changes in a matter of months. It encompasses the globe and shapes our planet, and yet can create intricate designs smaller than a pinhead. It is more inexorable than continental drift, more forceful than a volcano, more destructive than a tidal wave, and more inventive than us. Evolution is our creator. To know it is to know how and why we are here. To understand it is to have some of its awesome power under our own control.

UNDERSTANDING EVOLUTION—
THE GOOD, THE BAD, AND THE UGLY

We can see the power of evolution by its actions—the results of evolution are spectacular, and indeed they form the basis of all of the computing techniques described in this book. But the true power of evolution comes not from what it has done, but the way that it does it. In other words, before we can use evolution on our computers, we must explore the workings of evolution. Just how and why do things evolve?

This is a rather difficult question to answer in any rigorous sense, but (thankfully) quite easy to answer in an intuitive sense. But before we can investigate how evolution works, either rigorously or intuitively, we should clear up some confusions in the language. The word *evolution* is quite fashionable at present, being used to mean a surprising variety of things. We are confronted by "evolutionary cars" in advertisements, "evolutionary theories" by philosophers, and "evolutionary politics" by spin doctors. Planets are said to evolve, as are continents and even clouds. So that unnecessary confusions are avoided, in this book we must forgo such casual use of this important word.

Contrary to popular belief, evolution does not mean *better,* although it may be better than other processes at generating certain results. It does not mean *improvement,* although it can certainly produce dramatic improvements. Evolution does not mean *adapt* either, although animals evolved to fit specific niches are often called adapted. Indeed, as we shall see later, evolution does not even mean the *manipulation of genes.* All of these words describe aspects of evolution or results of evolution, but none captures the true essence of this fundamental process.

Evolution has a very specific and really quite simple meaning: It is a gradual process of directed change. But not just any such process can be called evolutionary. Three elements play the starring roles in an evolutionary process: the good, the bad, and the ugly.

THE GOOD

The good part about an evolutionary process is *having children.* If nothing ever reproduced, then evolution could never occur. Children form the engine of evolution.

To illustrate this, imagine that we are looking at a thing. Perhaps we have landed on the surface of another planet and are wondering about the origin of something that we see. That thing can be the result of evolution only if it (or others of its kind) is capable of reproduction. If the thing turns out to be a rock, then we can conclude that the rock did not evolve, for rocks cannot have children. But if the thing suddenly gives birth to a baby thing and that baby resembles its parent, then it is possible that the thing we saw did evolve.

To know for sure, we have to take a close look at the parent and child. We are looking for similarities. More specifically, we're looking for signs that the child has *inherited* some features of its parent. This is harder than it sounds. Imagine the thing is a rock that happened to split into two as we watched. Both parent and child do resemble each other, and the child has clearly inherited a large proportion of the parent's substance. How do we know that the rock is not some kind of reproducing thing, which creates offspring by fission? At this stage, it is impossible to tell. We must continue to watch and see what the child thing looks like when it becomes an adult. If it resembles its parent before the fission, then the chances are that it is not a rock after all but an entity capable of reproduction. And we know that the thing is an adult when it has children itself.

In fact, it is usually possible to determine inheritance with a little more certainty. Children do more than simply resemble their parents, they receive information transmitted from their parents. In nature, this information takes the form of fragments of DNA called *genes*. We know that a child has inherited features from its parents because we can locate in the child copies of the parents' genes that helped generate those features.

But evolution is not just about having children that resemble us. There is also a less pleasant, but equally necessary part of every evolutionary process.

THE BAD

The bad part of an evolutionary process is *selection*. In evolution, selection determines winners and losers. If you are a winner, then you may live a little longer and have a slightly better chance of having children. If you are a loser, at best you will have a reduced chance of having children; at worst, you will be dead.

Selection in nature normally involves much fighting and much death. It is a brutal mechanism that has little subtlety and is often deemed horrific in the light of the more benign beliefs and laws of our society. But however much we dislike the necessity for such suffering and death, without it we would not exist. It is only because of the countless failures and deaths that have preceded us over the last few billion years that you and I are here today.

Selection is the steering wheel of evolution. Without selection, everything would simply reproduce and die randomly. There would be no concept of some things being better able to survive than others. Such a state of affairs cannot happen in the real world—selection is built into the laws of physics. In our computers, however, we can turn off selection in our evolutionary programs. What we get is pure randomness.

Selection is not random. In nature, creatures are selected according to how well they can survive in their environment. A gazelle that is not as good as its companions at fleeing from a cheetah will die. A gazelle that is better than its companions at fleeing such threats will live a little longer than its companions. On average, gazelles with longer life spans have more opportunity to have offspring. These gazelles are winners. Nature also uses competition as a selection mechanism. A lion better able to drive off other lions and take over a pride will have more offspring than a less capable lion. The lion is a winner. And there are more subtle selection pressures as well. An animal more inclined to reproduce is more likely to have offspring than an animal less inclined toward such activities. An animal that helps to ensure that its offspring survive by providing food and protection is likely to have more healthy offspring than one that does less for its offspring. In some environments, a creature that produces thousands of offspring just so that one or two may survive will result in more progeny surviving than a creature that produces only hundreds of offspring. These are all methods of winning in selection.

If you are a good survivor and parent, you will grant your children a better chance of survival, whether it is because there are more of them or because they are simply healthier. The more successful your children are, the more likely it is that some of your features that they inherit will be passed on to future generations. So if you happen to be good at fleeing cheetahs, fighting other lions, or just having lots of children, then your offspring are likely to inherit these traits. In a nutshell, this is evolution. If you are good enough to be a winner in the harsh game of selection, then your children are likely to be equally good. If you are a loser, then your inferior traits are less likely to be passed on to future generations.

Selection clearly plays just as important a role as reproduction, but there is one other essential element to an evolutionary process. Without it, we would all be identical, and evolution could not occur.

THE UGLY

Mistakes, random shuffling, and mutations are ugly and disruptive processes that affect our genetic makeup. Nevertheless, they form the third vital component of evolution: *variation*.

Evolution needs variation in order to work. Our children must not be identical to ourselves. There must be some differences. Some must be short, some tall. A few must be more intelligent, a few less so. Some must be attractive, and yes, some must be unattractive.

Variation is caused by a number of mechanisms in natural evolution. Sexual reproduction causes the most commonly seen variation, as the traits from two parents are randomly shuffled together in novel combinations to form children. We all have experiences of this type of variation: "he has his mother's eyes and his father's ears." Random variation can also be caused by errors during the copying of our genes. Such errors result in altered, duplicated, or missing genes, which can cause new traits to emerge in offspring. Finally, variation can be caused by mutations in the genetic structure. If the copies of genes provided by either parent have been damaged by radiation (and this may be anything from normal background radiation to the high levels of man-made radiation caused by some technologies), then mutations will provide novel variation in children.

It is the last two types of variation that seem to provoke the most fear in prospective parents. The idea of mutant children evokes mental images of green-skinned monsters with six eyes and five arms. Although it is true that some mutations or genetic errors can cause severe disablement for individuals, this is rare.* Most mutations cause small changes to us,† and these changes are essential.

Variation is so important because selection requires difference. If everything were the same, then everything would be equally good or bad at surviving and producing offspring. Selection would have the same ef-

---

*The reason mutation generally does not produce large changes is also because of evolution. The rate of mutation and the degree to which a mutation will affect us has evolved as we have evolved. Most major mutations cause children to be less fit than their parents, so those individuals less likely to have (and be adversely affected by) such disruptive mutations tend to pass on this trait more frequently.

†I am an example of a "mutant"—I used to have an extra tooth between the lower two front teeth. This caused me no problems except a couple of trips to the dentist to have it removed.

fect on everything: either everything would live or everything would die. Variation allows selection to enhance some traits and reduce others. Whatever reproduces with variation and is operated on by selection must evolve. So the fact that we all contain a few novel mutations is good, not bad. If we all had the same level of intelligence, for example, there would be no possibility that our intelligence could increase in future generations (I'm assuming here that increased intelligence is beneficial).

Of course, these three processes are not really good, bad, or ugly. Such labels are helpful for explanatory purposes and make a useful memory aid, but they are a little emotive. Evolution requires the three elements: reproduction with inheritance, selection, and variation. These three are not good, bad, or anything else. Like evolution, they just simply *are*.

Having cleared that up (although there will no doubt be complaints anyway), let's proceed with this investigation into evolution. So far we have looked at how evolution works. It is time to tackle another fundamental question: *What* does evolution evolve?

THE UNIVERSALITY OF EVOLUTION: FROM GENES TO BITS

GENES

Our imaginary movie showed how life evolved on Earth, and we've now seen exactly how this evolutionary process happens. Clearly, evolution evolves organisms. Or does it? Actually, this is an oversimplification.

In nature, although the three elements of evolution do appear on the surface to affect organisms, in reality they operate at a much lower level. Somewhat counterintuitively, reproduction, selection, and variability affect only *genes* directly. The genes define how organisms will develop, so any changes to the genes will usually cause corresponding changes to the organism. This often provides the illusion that organisms are evolved. Also, because we are organisms ourselves and because genes often seem to be some kind of invisible fairy tale, most of us find it easier to think of evolution as manipulating whole organisms. However, we should try to avoid this fallacy. As embryologist Lewis Wolpert adamantly (and repeatedly) says, "The only thing that changes in evolution is the genes. *Nothing else."*

Let's look at each of the elements of evolution, to see how the genes are affected. It is often said that reproduction is one of the fundamental properties of life. Genes are not alive, organisms are. Surely, it is the living creatures that have offspring, not their genes. But appearances can be deceptive. When we reproduce, we do not provide our children with copies of ourselves, we provide them with copies of our genes. We all develop from a single fertilized egg, using strands of DNA containing genes provided by our parents. Any changes made to us during our lifetimes cannot be written into our genes. If I do a lot of weight training and build up some big muscles, this does not mean that my children will be born with similarly big muscles. However, if my genes specified that I should develop large muscles, then the chances are that some of my children would inherit copies of the same genes, and they would be equally muscular. Reproduction happens at the level of genes, not organisms.

The same argument can be used for variation. Imagine you have an accident that results in your losing your left thumb. You, an organism, have been affected by a random variation. However, when your children are born, they have two thumbs as normal. The variation has disappeared! So if variation of organisms is not inherited, then evolution must not be involved. Imagine instead that a collection of genes in one of your reproductive cells (sperm or egg) undergoes a mutation. This change results in your child being born without its left thumb. When he or she becomes adult and also has children, some of them are also born without left thumbs. Clearly, the variation is inherited, and so will play a part in evolution. Evolutionary variation affects genes, not organisms.

But what about selection? Surely organisms, not genes, survive in their environments and fight for mates? Surely it must be the better organisms that are selected? Again, this is an illusion. Imagine a cheetah that can run a little faster than the others and is a little stronger. It always has more success hunting, and so it is able to have more children, and these children are healthier. Our cheetah is a winner, and the traits that made it so are likely to be passed on to its children, and their children. How? Through its genes. The "run fast" genes and the "strong legs" genes were selected because of the success of the cheetah during its life.

Organisms provide the means of evaluating genes, but in the end, the genes are the only part of the organism that can be passed on, so only the genes can be selected in evolution. As geneticist Steve Jones says, "Evolution equals genetics plus time."

OTHER MEDIA

So evolution evolves genes. But it turns out that evolution also evolves other things. In fact, a surprising variety of things evolve, many having nothing at all to do with nature. This might seem strange to those unaccustomed to thinking about evolution in other media, but it need not be.

In some respects, evolution is a bit like the wind. Wind is an abstract concept; it implies nothing about the medium in which it operates. While we're used to windy air, wind can disturb any gas. When our space probes had a close look at our neighboring planets, they observed this. Jupiter's great red spot, for example, is one gigantic (and very windy) storm of ammonia. Likewise, Neptune's dark spot is a storm of methane. It is even possible to have windy liquids, although we normally refer to such effects as currents rather than winds.

In exactly the same way, evolution is a concept independent of medium. We are all used to the idea of evolution in nature as depicted by the imaginary film earlier, but this is only one medium in which evolution operates. Another is language.

Our languages are not static. Every year, new words are added to the dictionary, and the meanings of existing words change. Some words are borrowed and mutated from other languages, and others fall out of fashion and become disused. Our languages are evolving as we speak them, and they have been since their conception hundreds of thousands of years ago. Indeed, Charles Darwin was well aware of and influenced by theories of language evolution as he formulated his theory of natural selection.

Let's take a closer look to see whether languages truly evolve. A language, like an organism, is reproduced or spread by copying smaller subcomponents. Instead of genes, languages are defined by words, meanings, and rules of grammar. When a parent teaches a child language, that language is being reproduced in the mind of the child. The child will be given (inherit) the words and meanings of that language from its parent.

However, although both parent and child speak the same language, each has a subtly different version stored in his brain. Because of misunderstandings, mistakes, and other random errors, each of us often uses the same words to mean slightly different things.

Variation can also be caused by people creating new words or redefining the meaning of existing words. And selection of words, meanings, and grammar rules happens also. Words that help us to communicate are more useful than those that hinder this process. Most of us tend to favor words that are most comprehensible and use unintelligible words less. As we reproduce our own personalized version of our language in our children (and others of society we happen to communicate with), this word selection process becomes evident. Newcomers to the language learn only words that win in selection; the losing words slowly fade away.

Because the three processes of reproduction, selection, and variation are clearly evident in language, we can confidently conclude that our languages do evolve. Unlike natural evolution, which operates in the physical universe of atoms, molecules, chemicals, and physical laws, language evolution operates in a different medium. Our languages evolve in a conceptual universe made up of brains, books, speech, and writing—two entirely different universes and yet the same process of evolution.

BITS

In recent years another type of universe has been steadily growing in complexity. It is a universe of highs and lows, of 1s and 0s. I am referring to the digital universes in our computers.

In nature, a very special molecule made from genes, called DNA, becomes self-replicating because of the laws of physics. In our computers, we can write software that treats certain collections of bits in exactly the same way. These bits become binary DNA, capable of self-replication because of the software laws of physics. In nature, the laws of physics and the physical environment determine which genes are more beneficial than others. In our computers, software laws and environments can be created to judge the effectiveness of binary genes. In nature, the environment and mistakes in self-replication cause DNA to vary and change. In our computers, digital environments and randomized errors can be used to generate variation in binary genes.

In short, we can create software that allows bits to reproduce, to be selected, and to vary in the digital universes of our computers. These programs form the environment in which evolution can occur in our computers. Although natural evolution inspired these computational techniques, we are not talking about simulated or artificial evolution here. Just as natural evolution and language evolution both exist in different universes, computational evolution is a valid and real form of evolution, which can and does exist in our computers. The programs that permit things to evolve in our computers are called evolutionary algorithms.

EVOLUTIONARY ALGORITHMS

As I type on my portable computer, my feet are resting on a unique piece of furniture. There is no other piece of furniture quite like it in the world. This wooden structure is a world first. It is entirely original. It is my coffee table.

I know that my little table is unique because I know who designed it. Or to be more precise, I know who *didn't* design it. People didn't. My table was not designed by a human being.

So who did design it? Did some aliens from outer space land in my back garden and decide to design me a coffee table? Did elves at the North Pole carefully create this piece of furniture? No. Nobody designed my table. No conscious, thinking designer created the clever structure that supports my tired feet. The design of my table was *evolved*.

Some time ago, my computer ran a program. The software instructed the computer to create a digital universe in which solid objects could exist. The objects would fall over if they were unsteady, and if they did, they were considered to be unfit. Small particles were dropped onto the objects. If the particles were all supported at a certain height, the objects were considered to be fit. Other aspects such as mass and size also affected the fitness of the objects.

In this universe, every solid object was capable of reproduction: it could have offspring. Every so often, an object would find a partner and have a couple of children. The fitter objects had more success on average in producing offspring.

Every object in this universe also had its own genotype. This collection of binary genes defined the shape of the object—it described how the object should be made. When two objects had offspring, copies of their binary genotypes were mixed up. The children inherited fragments of the genotypes from both parents. They inherited their parents' genes. Once in a while, a very small part of a gene would change slightly, resulting in a slight change in the object when it developed.

Objects lived in this universe, trying not to fall over and trying to support particles at the right height while being fairly light and small. The better objects had more children. Every object had a life span, and should an object reach it, it would vanish from the universe. However, objects that were very unfit tended to disappear a little sooner, as pressure for space was exerted by the new children.

As this program ran, it was possible to see some of the objects as they "lived" in this cosmos. They didn't run around or sing any songs, but they did slowly change. At the beginning, all objects began life as random blobs. Generation by generation, objects stopped falling over. They started to develop flat upper surfaces to support those falling particles. They became small and lightweight. The objects became more and more like tables.

After five hundred generations of continuous evolution, the program ended, and the best object was stored to disk. The result: an original coffee table design (see plate 1). And this was no fluke. Every time the program is run, it produces an excellent coffee table design, and every time the design is slightly different.

GENETIC ALGORITHM

The program is an example of a genetic algorithm, or GA as the researchers in this field refer to it. It has this name because it consists of a set of rules that define how genetic material made from binary numbers should be manipulated in a digital universe.

Before we continue, I should say that the genetic algorithm is just one evolutionary algorithm. Others include evolutionary strategies, evolutionary programming, genetic programming, and memetic algorithms. All evolve solutions to problems, and although many were independently discovered, they all work in much the same way as the GA. Indeed, the only reason I mention the different algorithms is that some researchers prefer some to others and will become somewhat annoyed if I don't mention their favorite. Scientists can be very illogical and emotional, like the rest of the population!

Anyway, the GA I wrote to evolve my coffee table was quite complicated, but underneath the complexity, it followed the same fundamental rules. Let's look at a simpler problem and see how the genetic algorithm evolves solutions to it.

Many years ago I was given a Venus flytrap as a present. The watering instructions for the plant said, "Keep soil moist at all times," because it came from a swampy environment. I didn't want it to die because it was too dry, so I made sure it had plenty of water. Unfortunately, I overwatered the plant, and it died anyway.

This is an example of a problem with one parameter. The only thing we will change is how much water we give to the plant. Clearly my solution to the problem was wrong. But there must be a solution—an amount of water per day that will not kill the plant. Finding the solution is rather difficult, for even if we assume that we can duplicate the behavior of a Venus flytrap exactly in the computer, I would be just as likely to over- or underwater the digital plant as I would the natural one.

But evolution is rather good at finding solutions to difficult problems. And if we have created a digital plant in our computer, then we can use a genetic algorithm to evolve the solution to this problem. Since the solution is a specific number telling us how much water to give the plant per day, then this is what the GA will evolve. Instead of evolving an object such as my coffee table, the GA will evolve a number.

Perhaps surprisingly, the GA evolves a number in exactly the same way as before. It maintains a population of numbers. The fitness of each number is calculated. (For this problem, the longer the plant lives, the better the number is.) The fitter numbers find partners and have children. Every number has a corresponding gene. The children inherit random chunks of their parents' genes. Having had children, the parents disappear, and the population is filled with the new numbers. These numbers are also evaluated to see how fit they are, the best have more children, and so on.

Because this is a simple problem, we can look at the genetics of the solutions. The genetic algorithm usually employs *binary genetic encoding*.

This scary-sounding terminology simply means that a binary value is used as the gene for each number. So if one of the numbers being evolved by the GA was 13, its gene would be 00001101 (which is 13 in base two, or binary). If another of the numbers was 156, then its gene would be 10011100. And if these two numbers were fit enough to have offspring, then the GA would use crossover and mutation to chop up and vary these two genes and make two children. Again, it's a surprisingly simple procedure. A random crossover point is chosen, say, 5. The first child then receives the first 5 bits from parent 1 and the remaining bits from parent 2. The second child receives the first 5 bits from parent 2 and the remaining bits from parent 1. So if our parents were 13 and 156 and the crossover point was 5, then the children would have the genes 00001100 and 10011101. Mutation is also occasionally applied by GAs. This is even easier to follow. A random bit is selected, and its value is flipped. If the first child was mutated, the random point might be 7. Its gene would then be mutated from 00001100 to 00001110.

These genes define the number for each child. If the first child had 00001110 as its gene, then its number would be 14. If the second had 10011101 as its gene, its number would be 157. So the two parents, 13 and 156, had two children, 14 and 157. Because the children inherited their parents' genes, they resemble their parents.

Perhaps surprisingly, it is normal practice to begin evolution from a population of entirely random genes. The GA then has the chance to explore far more potential solutions. Because only the fitter individuals have offspring, evolution quickly discards the poor solutions and homes in on the best solutions.

Using these simple binary-shuffling mechanisms, the genetic algorithm would evolve a good solution to the Venus flytrap watering problem. And as long as the behavior of the digital plant matched that of the natural one, I would stand a much better chance of keeping the flytrap alive using the evolved solution.

USING EVOLUTIONARY COMPUTATION

Clearly there are more difficult and interesting problems than the task of preventing me from killing my household plants. Nevertheless, this slightly daft example does show something rather important: how evolution works in computers. Whether we are dealing with numbers, symbols, or coffee tables, we can create a suitable digital universe that will allow these things to reproduce with inheritance and variation and that will have some form of selection. When we enable these three processes to occur, evolution simply happens. And when evolution happens in a computer, the results are often surprising.

We have harnessed *the* force of nature, and it does sometimes feel as though we are riding a tornado, barely in control. I use such colorful language because we cannot predict exactly what evolution is going to do next. Even after more than thirty years of analysis by some of the best minds in the world, we have no useful mathematical proofs for evolution. We cannot *prove* that evolution will even find us a good solution—but it almost invariably does. And we certainly cannot predict the solutions that evolution generates. Even for the extremely simple Venus flytrap problem, where we were evolving just one number with a genetic algorithm, we have only a few unproven theories on why it works at all.

Perhaps unsurprisingly, some researchers find the whole thing rather scary. Evolution is a wild, powerful beast, tolerating us, but remaining unpredictable and untamed in our digital universes.

The lack of mathematical proofs is more than enough to put many people off evolution altogether. How can we trust something that is so unpredictable? But increasingly the benefits of evolution are outweighing such concerns because evolution works rather well. Although we can't prove too much, the more we use evolution to solve our problems, the more we discover just how good it is at finding fit solutions. It is now clear that evolution is one of the very best techniques we have in computer science—it regularly outperforms traditional methods that were carefully created by researchers. The power of evolution so evident in nature is equally evident in the digital universes of our computers.

Even the oft-quoted misconceptions that evolution must take a vast amount of time or be very inefficient do not hold water. Today's computers allow us to evolve solutions for even the most complex problems in seconds or, at worst, minutes. In contrast, many of our carefully designed formal approaches can take considerably longer and will often be incapable of matching the quality of solution provided by evolution. Just recently, someone tried to argue with me that evolution is inefficient because of the number of solutions that must be considered before the best is evolved. This is silly. You might as well argue that it is inefficient for scientists to consider different theories as they try to understand something. If we didn't do this, our solutions would be pretty shoddy. Likewise, if evolution didn't consider different solutions along the way, it would never find the final solution. Just as we cannot jump from the Stone Age to the Space Age in a single giant leap, no technique can jump straight to a good solution for these difficult problems. Evolution is actually one of the most efficient ways we know of finding good solutions for many different problems.

The number of different applications that we have successfully used evolution for is immense. Indeed, I fear that the list is so long that I would only bore you with the details. So I will not tell you about the jet engine turbine blades that have been evolved, or the aerodynamic cars, or the satellite structures, the flywheels, the optical prisms, the photorealistic faces, the factory schedules, the school timetables, the fraud-detection systems, or the architecture that have all been evolved. I won't mention the way evolution has been used to enable data to be represented using facial expressions, or the evolved game-playing strategies that can easily beat humans, or the evolved fighter-plane maneuvers that allow our aircraft to survive dogfights. Instead we shall concentrate on just two applications for the remainder of this chapter. (Nevertheless, this chapter is not the only one to talk about evolutionary computation. Most of the other forms of digital biology described in following chapters have at least part of their design evolved, so you will be reading more on this subject later.)

Let's start with one of the newcomers to the field of evolutionary computation, genetic programming.

GENETIC PROGRAMMING

It may sound like something written by science-fiction author Arthur C. Clarke, but the aim of hundreds of researchers is to enable computers to program themselves.

Imagine it. No more programmers, no frustrations of waiting for the next version of software or for patches to cure existing bugs. Just tell the computer what you want it to do, and it programs itself. If your computer suffered from the millennium bug problem, all that you would have needed to do was say to the computer, "I'd like this software to work after the year 1999," and it would have reprogrammed itself. This is *automatic programming.*

Unfortunately, as you might expect, making a computer program itself is a rather difficult problem. Just as we can say an infinite number of things in our language, there are an infinite number of computer programs that can be expressed using a typical programming language. Trying to work out which one to use is like looking for a needle in an infinitely big haystack.

Luckily, we know of a handy approach that copes with difficult problems such as these: evolution. By using a slightly modified genetic algorithm, we can evolve computer programs. The algorithm is called *genetic programming,* and an entire research field has grown up to do just this.

In genetic programming, or GP as it is usually called, evolution is used to modify program instructions. In the example in the previous chapter, simple instructions such as *add, store,* and *output* were used in an example program. GP evolves instructions such as these. It uses evolution to put the right set of instructions together, which can then be run and will make the computer do the desired action. In a sense, it is like evolution in one digital universe automatically creating another digital universe.

This is a very hard problem. Even evolution struggles, and the GP algorithm suffers from a few problems that may hinder progress. So far, researchers have managed to allow the evolution of simple programs that contain mathematical expressions, and the first programs with loops and function calls are now evolving. Nevertheless, GP works well enough to generate some impressive results. John Koza, the GP guru at Stanford University, has so much confidence in this approach that he has had his company, Genetic Programming Inc., build a 1,000-processor supercomputer for the sole purpose of evolving solutions to problems. This is one of the fastest computers in the world, and John's ambitions for GP are equally high. In his own words, "We are currently working in the areas of automated synthesis of analog electrical circuits and controllers, problems in computational molecular biology, various other problems involving cellular automata, multi-agent systems, operations research, and other areas of design, and using genetic programming as an auto-mated 'invention machine' for creating new and useful patentable inventions." John likes to patent things, indeed, he has patented GP itself. The rest of the research community is looking on with considerable interest to see which evolved invention will be the first to be patented also.

John Koza is not the only one to achieve success with GP. I will tell you about others in Chapter 8, on growth, but for now let's concentrate on art.

Computer graphics can be used to produce some stunning special effects in films. Most book jackets and images in magazines and newspapers are produced or manipulated by computers. So clearly, computers can be used to generate some amazing images. Some artists have taken this idea a little further. What if you had a computer program that generated a piece of art? And how do we write that program?

It is the second question that causes the problems. Working out which instructions to use in order to tell the computer to "display something artistic" is not easy. And even if you manage it once, you'll still have to write a new program (or modify the existing one) for every new piece of computer art you desire. It seems like a lot of work; if only we could say to the computer, "Write me a program that generates a new piece of art."

The natural answer seemed to be GP. And numerous artists around the world have used GP or permutations of it to evolve programs that produce artistic images. Karl Sims at MIT was one of the first to do this. In the early 1990s, just after the birth of GP, Karl developed a system that evolved programs that displayed remarkable two-dimensional images. To work out how fit each program was, their images were shown to the user, who would judge them: "I like this one more than that one." The programs that produced the nicer images would then have more off-spring, and over a number of generations, evolution produced programs capable of displaying what can only be described as art. Because of the unpredictable nature of evolution, every time it was used, a different program was evolved and a different piece of art was generated. Karl had created an automatic art generator.

As is usual in the arts, controversy is the norm. Computer art is still a widely debated topic, with many refusing to acknowledge any form of evolved image as art. Nevertheless, the idea of an automatic art generator has become increasingly popular over the past few years. We may not all be artists, but most of us know what we like and what we don't like—and that's all that is necessary for evolution to generate a personalized, unique picture for us.

Karl was not the first to evolve art. William Latham, formerly of the Royal College of Art, teamed up with IBM research scientist Stephen Todd and evolved stunning two- and three-dimensional pieces using a different evolutionary algorithm in the late 1980s. Steven Rooke is one of the more recent evolutionary artists to emerge. Steven is an extremely enthusiastic champion of evolutionary art. To date, he has spent seven years evolving ever more evocative images. His passion for this subject is evident through his writing: "I can't stop. There is something compelling about this process. It feels as though the images are *trying* to break out of their hyperspace into the physical world. . . . I do see these images in dreams, in altered states. When I hear an audience gasp during a slide show or multimedia presentation, I have to think there is something broader happening, beyond the whims of an eccentric arti-scientist. Time and the marketplace will tell."*

Steven's belief in evolutionary art is borne out by the images that are generated (see plate 2). While not everyone would want them hung on their walls, most would agree that they are impressive accomplishments. The underlying programs that evolve to display such images are also impressive and far too complex to allow real comprehension of their workings. In a competition between a human-designed program and an evolved program to generate the best art, there would be no contest: evolution would win hands down.

EVOLVABLE HARDWARE

Enabling computers to program themselves is just one of the success stories for this type of digital biology. Another very exciting area is the use of computers to design electronic circuits and even other computers.

Evolution has been used to improve or generate new designs for some years. Indeed, evolutionary design is one of my own pet subjects. (Don't

---

*From the on-line biographical notes of Steven Rooke.

worry, I won't bore you with too many details of that.) The use of evolution for electronic circuit design has become so popular that it has now developed into a research field with its own name: *evolvable hardware.*

The first exploits of researchers in this field were relatively simple. Placing the microscopic components on silicon chips had always been rather tricky for electronics engineers. How best to use all of the silicon without waste? The first applications of evolution for this problem were to design sensible organizations of components and collections of components. It seemed to work quite well but was never a terribly earth-shattering achievement. However, the field quickly moved on from these humble beginnings. Researchers such as Adrian Thompson at Sussex University began showing what evolution could really do with electronics.

Adrian allowed evolution to do something rather clever: he allowed it to evolve real electronic circuits. Using a new technology called *field programmable gate arrays* (FPGAs), he was able to evolve new circuit designs in the computer and then test them in real silicon immediately. We won't go into the gory details of how FPGAs work—all that you need to know is that they can be reconfigured to act like any circuit by applying electrical signals to them. So instead of having to get each new circuit manufactured in a costly silicon-chip-producing factory, a single FPGA can be instantly reconfigured into a new circuit design.

Adrian gave evolution quite a lot of freedom. He simply said to the computer, "I want a circuit that does X" and let evolution generate any circuit it wanted to using the FPGA. The results caused quite a stir in the research community. Circuits that behaved very bizarrely kept emerging. They performed the desired function, but it wasn't at all clear how they worked. When Adrian had a closer look, he discovered something fascinating.

Digital electronics uses tiny components to switch the flow of electricity on and off within circuits. When a human electronics engineer designs a digital circuit, he assumes that the components do exactly that. In reality, however, the tiny components are not perfect—there will be unimaginably small periods of time while a component is switching from on to off or off to on, when the electricity is neither fully on nor fully off. We don't know enough about the characteristics of the silicon used for each chip to be able to use this in-between behavior when we design circuits. But evolution does.

Because evolution tested its circuits on a real piece of silicon, Adrian discovered that it quickly learned to take advantage of the tiny flaws that caused those different in-between switching states. Evolution built entire circuits using these characteristics, making digital circuits that behaved as analog circuits, oscillating and doing extremely strange things in order to make the circuit perform the desired function. The circuits were nonhuman in their design, they were very compact, efficient, and worked using principles that human designers would never dream of employing.

In fact, evolution was so good at using the specific characteristics of each chip to make the circuits work that Adrian discovered that the same circuit wouldn't work on any other chips. This problem has now been overcome by testing on multiple chips during evolution. Even so, many of the evolved circuits still defy analysis. Adrian is even resorting to measuring the temperature of the chips to try to work out which parts are being used by the circuits as they run—rather like neurologists using brain scanners to find out how the brain works.

This is not the only example of evolvable hardware giving surprising results. Julian Miller, now at Birmingham University, has also performed research on evolving circuit designs, although these designs are tested in the digital universe of his computer and not in physical silicon. Julian has shown how unconventional circuits are evolved for a number of applications. The circuits step out of the constraints imposed by traditional mathematical algebras used to design solutions. Because evolution does not know or care about our conventional wisdom and our theories of circuit design, it ignores them. Instead, evolution generates novel and very efficient new circuits—designs that we may learn from. As Julian says in one of his research papers, "[I] feel confident that the process of learning new principles from a blind evolutionary process is inevitable, it is just a matter of time."

Evolvable hardware continues to expand in scope, just as the whole field of evolutionary computation does. Researchers such as Moshe Sipper at the Swiss Federal Institute of Technology are currently investigating circuits that can reconfigure themselves should they be damaged.

With NASA and other research-and-development organizations around the world becoming involved, it seems likely that evolution will allow self-repairing circuits to become a reality in the next five to ten years. To illustrate the kinds of things researchers are working on right now, just take a look at the list of topics in a recent conference on evolvable hardware. Most would fit seamlessly into any *Star Trek* episode:*

Third International Conference on Evolvable Systems:
From Biology to Hardware

TOPICS (to be covered include, but not be limited to)

- Evolving hardware systems
- Evolutionary hardware design methodologies
- Evolutionary design of electronic circuits
- Co-evolution of hybrid systems
- Hardware/software co-evolution
- Intrinsic and on-line evolution
- Evolutionary algorithm implementation in hardware
- Reconfigurable hardware
- Self-replicating hardware
- Self-repairing hardware
- Neural hardware
- Adaptive hardware platforms
- Bio-robotics
- Applications of nanotechnology
- Biological- and chemical-based systems
- DNA computing
- Evolving controllers
- Real-world applications of evolvable hardware

The capabilities of evolution within the digital universes of our computers seem to increase daily. We are now beginning to understand the kinds of things evolution is best at and how to make it achieve those

---

*If some of these sound just too interesting to ignore, don't worry—I'll be describing most in the following chapters. Alternatively, if these sound rather dull, don't worry—I'll be describing lots of other things, too.

things. And when we use evolution to help create other forms of digital biology, we increase the capabilities of computers still further.

SUMMARY

As we saw in the imaginary movie at the beginning, natural evolution is astonishingly powerful and diverse in its creations, and it all stems from three deceptively simple processes: inheritance, selection, and variation. These act on our genes to change life on our planet continuously.

Evolution affects more than genes. We can see evolutionary processes happening in our languages, as our words evolve, and maybe in our minds, as ideas evolve. And when inheritance, selection, and variation occur to bits within our computers, those bits also evolve.

From coffee tables to Venus flytraps, evolutionary computation is a force to be reckoned with. Computers now program themselves by evolving programs, those evolved programs can generate art, and computers can design and even build highly original electronic circuit designs.

In a very real sense, evolution is our creator. In an equally real sense, evolution is now enabling us to transform our technology.

The next chapter describes one of evolution's masterpieces of design: brains.

BRAINS

"AND, FOR EXAMPLE, we've been doing a lot of work with Sarah-Jayne Blakemore on *tickling*," continued Professor Chris Frith. "Because the idea is, the reason you can't tickle yourself is because you predict the state and it cancels out. . . . I mean, this has been known for thousands of years, no doubt, but was proved by Larry Weiskrantz in 1971 or something. But it is indeed the case that if you ask people to rate the intensity of the sensation, it's much less if you do it than if somebody else makes exactly the same movement. So she did some very nice experiments on the imaging side where she used robots to do the tickling."

I had gone to see Professor Frith on a stormy May afternoon to talk about the brain and consciousness. Chris is a bit of a world expert, particularly on matters to do with mental disorders and how they affect our sense of self. To find the conversation straying onto the subject of tickling was a little unexpected.

"If you look at the *brain activity*," he said excitedly, presenting me with a picture of a brain with two little glowing spots on it. "This is the so-called secondary somatosensory cortex, where . . . anyway, it's one of the first areas in the cortex where somatosensory stimuli are processed." He then dug out another piece of paper with a little bar chart of brain activity. "And this is just showing . . . this is the activity in the regions when somebody is tickling you, compared to you tickling yourself. . . . No, sorry, that one, yes, that's you tickling yourself. That's you moving, but not tickling yourself, because obviously you have to parcel those out, and that's just rest. So when you tickle yourself, there's a dramatic drop in brain activity—I mean it virtually disappears—no more than it is if you're just moving. But she then went on to look at the (I don't know if this is what you want to know about, but . . .) test more directly this prediction business, so she used these robots. So you move one robot with this hand," he said as he waved his left hand, "which feeds the signal to a second robot, which then tickles you. And then you can distort it. And basically, what she found was that"—he paused for breath—"if it's just the robot making movements, you feel highly ticklish. If it's directed by you, just like a rigid rod, the ticklishness goes right down. But then if you introduce, say, a hundred-millisecond delay, the tickling goes right up again. It does seem to depend on this quite precise prediction."

Chris leaned back in his chair and continued. "The reason for doing all this was my interest in schizophrenia. One of the symptoms that patients report is that when they move, it feels to them as if it's not them that's initiating the movement but some external force—a sort of delusion of control. So we said that maybe something is going wrong with the system, so they don't predict it properly. And our theory was that they should be able to tickle themselves, which apparently they can. So when you contrast them doing this and somebody else doing this, they don't make the distinction."

"So they can tickle themselves!" I said with a smile, as Chris chuckled gleefully. "But are these kinds of presumably different areas of the brain that in schizophrenic patients have become disconnected?"

"Well, that's the idea, yes," Chris replied. "That's what we'd like to show next, as it were. But a lot of the interesting imaging is because cognitive psychologists had models like this with little boxes in them"—Chris pointed to another piece of paper—"and one is very tempted to say, 'Ah-ha, can we find which bit of the brain corresponds to that box?' That's *mostly* what we do, although it's becoming slightly more sophisticated, but that's the essence of it."

I suspected that Professor Chris Frith was being overmodest about the work being performed by cognitive neurologists, but did not argue. Indeed, after I had interviewed psychologists, neurologists, neuroscientists, computer scientists, and philosophers, it had become very clear that the amount of research being performed on understanding the brain is vast. You could write a few hundred books on what we already know about the brain (and researchers have done exactly that). Using experiments such as the "tickle test" and many other equally bizarre ones, thousands of people worldwide have been working for years to uncover the mysteries locked inside each of our heads.

And there are a lot of mysteries to uncover. Our brains are the most impressive achievement of evolution we know of. They are several million times more powerful than any computer we've created. Every one of our brains is so complicated that it would take thousands of our best experts many lifetimes to design and construct anything a fraction as powerful using our current technology. Indeed, the technology in our heads is millions of years ahead of us; we are still like cave dwellers looking up at the sun in awe.

Our brains also create our sense of self, our consciousness. This very personal part of each of us is created by cells in our skull, a concept so hard to accept that a large percentage of the population instead believe that our "self" is generated by a supernatural soul.

Whatever you choose to believe, we're a long way from having all the answers about the brain. Psychologists tell us what brains do, neurologists tell us how they are organized, and neuroscientists tell us what they're made of, but nobody seems to know exactly how it all works. Only the philosophers have attempted to unify these seemingly unrelated areas. And as philosophers, their attempts seem to cause many lively debates, but agreement seems very hard to come by.

Nevertheless, despite the ongoing arguments, we know enough to understand the rudiments of decision making, learning, memory, and even consciousness. In this chapter, we look at what we do know about the brain. As you will see, we certainly know enough to be able to build digital brains within our computers that can make decisions, learn, and remember.

But for now let's put tickle tests and digital brains to one side. Before we can understand either, we need to look at the smallest components of our brains: the things that cause all those wavy lines on an electroencephalogram (EEG) display, the little cogs that turn when we think, the stuff that makes gray matter gray.

GRAY MATTER

Like it or not—and I certainly don't like it one bit—our brains are physical things. Our movements, behavior, memory, emotions, motives, ideas, decisions, perceptions, thoughts, and personalities are generated by a physical organ. If there is such a thing as a soul, it does not do any of these things.

"How can he be so sure?" you may be thinking. I am so sure for a good reason: we may not know how the brain works, but we certainly know what happens when it doesn't work properly.

Over the years, there have been thousands of gruesome head injuries. Indeed, as you read this, it is a near certainty that someone, somewhere in the world is going through the frightening experience of a car crash, fall, or stroke resulting in brain damage. Our doctors and scientists are used to seeing people with large chunks of their brains missing or dead. And depending on which chunk of the brain no longer functions, the patient will show marked changes. People lose their memories or sometimes just very specific types of memories—perhaps the names of animals. They lose the ability to move their limbs or speak. They may start behaving very strangely, become paranoid, feel as though God is speaking to them, or be unable to make decisions. They may become unable to comprehend anything on their left side. Even that unique part of us that we treasure so much, our personality, can become changed beyond recognition when brain damage occurs.

All rather depressing, perhaps, but extremely useful for understanding ourselves. We are physical creatures, and, like the other creatures on this planet, the brain is the control center, the "guy in charge of the body." And the brain is, in the end, just another organ in our bodies. Like the lungs, the heart, or the skin, the brain consists of lots of cells grouped together that are specialized to do a certain job. They all need oxygen and food, which is provided by a blood supply. They are prone to infection, and they like to be kept at the right temperature.

But as you might expect, brain cells are a bit special. First, they are all encased in a protective bone shell (your skull) to keep them from harm. They also don't regenerate terribly well. You have almost your entire quota of brain cells when you are born—one reason that babies' heads are so big and giving birth is so painful. By the age of two, your brain is about as big as it will ever get, and it's downhill from there, with brain cells dying every day without being replaced.* But thankfully we all have quite a few brain cells. Around 100 billion, in fact. That's around 15 times the number of people living on this planet—an impressive crowd packed in your head.

Brain cells are also special in other ways. They use electricity, for one thing. Just as we talk to each other on the telephone by sending electrical impulses down wires, brain cells talk to each other by sending electrical impulses down wires called axons. And just as we needed to run enormously long cables under the sea to connect continents together (before we had satellite communication systems), our bodies need to run long axons from the brain to muscles and senses. In a creature such as a blue whale, you can imagine just how long an axon linking a brain cell to the tip of its tail needs to be.

But if we're going to get to know the brain cell,† let me introduce you to it properly. You have 100 billion much the same inside your head, so you should try to remember its name. Meet the neuron.

NEURONS

There are as many as a hundred different types of neurons, so I hope you will forgive me if we explore only the most common features found in most neurons.

Your average, run-of-the-mill, standard kind of neuron (not that there is such a thing) has three important bits to it. First, like all cells, it has a cell body, which contains the nucleus with your DNA. Second, it has a long cable coming out of the cell body, called an axon. Finally, a neuron has an extensive "root system" of many branching cables, called dendrites, that go into the cell body. So your average neuron looks a little like the root system of a plant growing from a bulb (see plate 3). The

---

*Recent research has shown that new brain cells do seem to grow back in one region of the brain known as the hippocampus, a structure that plays a part in memory. Unfortunately, they don't seem to grow back anywhere else in the brain.

†There are actually many different types of brain cells—for example, glial cells, which help support and feed the other cells. Here I am referring to the cells that we use to think with.

roots stretching in all directions are the dendrites, the main stem is the axon, and the bulb is the cell body.

That's what it looks like. What it does is a little more complicated. Neurons use chemicals to generate electricity, which generates other chemicals, which generate electricity, and so on. It's a complex and carefully balanced cauldron of different substances that turn a cell into a simple computer.

To understand the working of a neuron a little more, think about something else for a while. Bear with me, for it does make sense in the end. Imagine you have a wineglass in front of you. Someone on your right is using a funnel to pour wine through a straw into your glass. The level of wine is slowly increasing. And someone on your left is drinking the wine from the glass through another straw, so the level of wine stops increasing.

The wine is red, and we all know that red wine stains are impossible to remove, so you are keeping a close eye on how full your glass is getting. Now imagine that the guy on your right is using a fatter straw. The wine is entering the glass faster than the person on your left can drink it. The level rises until—disaster—it overflows. Because you're now covered in red wine, you jump up and shout.

This strange little tale shows how a neuron works. The wineglass was performing one of the tasks of the cell body. The straws were acting as dendrites. You were acting as the axon, and the wine took the role of electricity.

Not convinced? Let me tell the same story, this time with the players unmasked. Imagine you have the cell body of a neuron in front of you. A dendrite is entering from the right. We'll call its signal excitatory (if you recall, it got you excited when the red wine spilled). A dendrite is entering from the left; we'll call its signal inhibitory. Electricity, a little like the wine, flows through both the dendrites, into the cell body. While there is a balance, the cell does not do very much. Now imagine that the electrical signal from the right dendrite is stronger than the one in the left. The electricity in the cell body quickly builds up until it reaches a certain threshold. When the threshold is reached, a pulse of electricity is sent down the axon.

So neurons send out pulses of electricity when excited beyond a specific threshold (around -0.055 volt). They are excited when they have more excitatory stimulation than inhibitory stimulation from dendrites. And the level of stimulation from dendrites is adjustable (through chemical changes rather than straw thicknesses).

This behavior makes each neuron like a simple computer. It sends out a binary signal depending on whether the combined inputs go above a threshold value or not. If this still sounds scary, just remember the wineglass. You sent out your signal (a cry of dismay) when the sum of wine entering and leaving the glass resulted in the wine overflowing.

So this is what a neuron does.* And the source of the electricity that tickles each of our neurons is, perhaps not surprisingly, other neurons.

NEURAL NETWORKS

Right now, your eyes are generating electricity. At the back of your eyeballs, your retina is constructed from thousands of tiny cells that convert light into countless pulses of electrical energy. These cells are types of neurons, and their signals run down axons that are gathered together in a place called the optic nerve, and meet dendrites of other neurons in your brain.† Those neurons may or may not fire pulses in response to the signals, but if they do, then their electrical pulses travel down their axons and meet dendrites of other neurons. Which may fire more pulses, which reach other neurons, which may fire more pulses, and so on.

The same thing happens for all of our other senses. Chemicals in our mouths are converted into the brain's electrical Morse code. Smell, touch, sound, heat, pain, balance, positional sensors in our limbs, sensors that tell us when our bladders are full or stomachs empty—all converted to electrical signals by cells and sent through axons (often via the spinal cord) to the other neurons in our brains.

---

*This has been a gross simplification of the workings of a neuron, but for the sake of brevity we shall not look at things such as timings, ion concentrations, and semipermeable membranes. Although they explain how a neuron generates an action potential in response to stimulation, such matters do not really help us to understand how we think.

†Actually, the signal travels through a number of neurons before even reaching the optic nerve, but I'm sure you get the idea.

Electricity in—and also electricity out. Electrical signals are sent down axons to muscles all over the body, causing us to breathe, move, shout, run, and everything else we do.

Our bodies contain the most complicated electrical systems that have ever existed on this planet. The wiring of your nervous system is more complicated than all the computers, the Internet, and indeed every electrical item we've ever made put together.

To understand how this can be true, remember that wineglass. It had two straws in it, corresponding to two dendrites for the cell body. The average neuron inside your head has many more than that. Because there are many dendrites and because dendrites branch out into many smaller and smaller rootlike structures, an average neuron is connected to 10,000 other neurons. That's quite a lot. Now remember that you have around 100 billion neurons inside your head. If you can imagine that each is connected to an average of 10,000 of its companions, you might begin to realize just how complicated your brain is.

So although a single neuron is capable of quite simple computations, when you have enough of them and when you connect enough of them together, you result in sufficient complexity for mysterious things like learning, memory, and even consciousness to emerge. It is the network of neurons that enables us to think.

WIRING IN YOUR HEAD

But if connections are everything, how does the brain know which neurons should be connected? We'll see later on that there are many specialized regions in the brain—areas that handle vision, speech, memory, smell, and so on. How do the neurons know to link the eyes to the vision centers, or the tickle-sensitive touch cells to the somatosensory cortex? How does the axon of a neuron for a muscle in your little toe find its way down via the spinal column from the brain?

Most of the work happens in the first two or three years of your existence. While still in the womb, the developing neurons in your brain send out tendrils. In the same way a shoot from a plant will seek out the light, axons seek out chemical signals that direct them toward the right areas of the body. The neurons growing in the retinas are programmed to seek out the visual centers of the brain, growing like a bloodhound following a scent.

To connect neurons to each other within the brain, some equally clever techniques are used. Shortly after birth, the number of synapses (links between dendrites and axons) increases dramatically—from 2,500 per neuron up to 18,000. The brain then spends the next few years doing lots of pruning. If connections between neurons are not being used, they fade away. It has been estimated that by the age of two, your brain has only 60 percent of the connections it once had. So the brain uses the method of "wire as many things together as possible, then remove the links that don't do anything." It's a little like building a bridge across a valley by filling the valley with scaffolding, then removing all the pieces that don't support anything. The result is an efficient bridge—or indeed, an efficient brain.

It has been suggested recently that when we are in challenging or stimulating new environments, our brains suddenly increase the number of connections again, before slowly pruning away the unnecessary ones in later months. Indeed, some researchers have even proposed that neurons themselves might be pruned away to help the wiring. It may be that the tiredness and slightly dazed feeling you get after your first few days in a new job may be hinting at some considerable work going on inside your brain.

The brain uses many other tricks to get its complicated wiring right. Some researchers say, "Neurons that fire together wire together." For example, when still only a few months old, our neurons must complete the wiring that enables us to see. Normally, with light entering both eyes, many neurons are stimulated simultaneously, and these form connections between each other, wiring the eyes into the brain. But when an eye infection causes the temporary loss of sight in one eye during this critical period, the neurons are not stimulated normally. Instead of wiring both eyes to the brain, all the "vision neurons" are wired to the single working eye. So even when the eye infection clears up, allowing the eye to see, because the neurons are not connected, the person will never be able to use the eye.

Normally, however, the intricate ways that the brain uses to keep itself in control and happy seem to work very well. They allow us to learn, remember, and adjust to new situations.

This brings us to the last major question about our gray matter. Exactly how does a cleverly wired network of neurons work? What's going on with all of those little neurons? How can they learn, remember, or do anything?

THE BUSINESS OF THOUGHT

Your Brain, Inc., is a company. It has a large number of employees and no single boss in control. But it does have a strict hierarchy of management, and it does believe in democracy.

The company is a consultancy. It doesn't make widgets; it analyzes data and provides information about what to do with it. Let's look at how it works for a fairly typical job.

One day the clients of *Your Brain, Inc.,* give the company an image. The clients have no idea what the image means or if anything useful is contained within it. The company gets straight to work, its image specialists first on the scene. Every "image employee" concentrates on a tiny part of the picture, and each reports to his immediate boss: "I think that this bit is dark" and "I think this bit is light" and "I think this bit is red." The bosses listen to information from their teams and accept the majority view (taking into account the fact that some of the employees are more reliable than others when talking to them). The bosses then report to their immediate bosses: "I think that this little area of the image may be dark and red" and "I think the area to the left is also dark and red." These bosses listen to information from their teams and also accept the majority view (again remembering who is usually most reliable). They then report to their bosses: "I think there may be a dark red line over there in the image" and "I think there may be another dark red line below." Their bosses listen to information from their teams, accept the majority view, and report to their bosses: "I think there may be a dark red rectangle over there" and "I think there may be a gray circle down there." Their bosses listen to the information and report to their bosses: "I think this looks a bit like a red sports car" and "I've just seen something similar, and it was farther away." Their bosses listen to the information and report, "I think there's a red sports car moving toward you." Their bosses listen and report: "There's a red sports car that is going to hit you if you don't move." Their bosses listen and report: "I suggest you get out of the way!"

Much the same process then happens with the decision about how to get out of the way of the sports car. Some employees suggest jumping, some suggest walking, some suggest standing still and shouting "Stop!" and some suggest running. The bosses, knowing who has given the most reliable suggestions in the past, decide on jumping. They then send a message to their clients: "Jump to the left."

Sometime later, many of the employees involved in the job hear from other staff that it went very well. The bosses take note and remember which members of their teams provided the right suggestions, so that in the future those employees become a little more influential when making suggestions. And because the employees who are regarded as most reliable are listened to the most, the company is constantly able to improve its overall ability to find the right solution more often and faster.

I'm sure I don't need to tell you that what you've just read is a description of your brain working as it processes something that you see (hence the name *Your Brain, Inc.*). The clients of the company were your senses and muscles. The employees and many layers of bosses were neurons.

What we just saw the company do is something that our brains do effortlessly: process visual information and use it to keep ourselves from harm. But the amount of computation needed for such a task is immense. Even our best supercomputers would struggle to perform the job of recognizing a dangerous car and moving the observer out of the way. Working out what things are and what they mean is extremely hard.

Neurons are exceptionally good at this kind of thing. (Indeed, in computer science when we want our computers to perform these types of task, we use digital neurons made from software to do it.) There are a number of good reasons that neurons are able to perform such impressive computations. First, the neurons are wired together in a hierarchical network. As we saw, this means that the first neurons do very simple things, like identify tiny patches of light. The information is then passed up the hierarchy through the network, where the data are summarized as "areas of light," then a line, then a shape, then an identifiable object, and so on. So although each neuron only ever processes very simple signals, the information those signals represent becomes progressively more summarized. And because the neurons are networked, they integrate the current summary with other information provided by other neurons in the form of memory or sensory information.

The hierarchical network structure of our brains allows simple neurons to do extremely clever things. It also makes our brains *massively parallel*. Unlike a normal computer, which must carry out its instructions one at a time, the billions of neurons in our heads can fire at the same time. So while some are working out that we're seeing a sports car, others are working out that it's moving, and others are thinking, "The driver looks distracted." Many more are busily making your heart and lungs work, keeping you standing upright, making your head turn and your eyes move. The chances are that there are even more in the process of working out what you're planning to buy from the shops even as you jump out of the path of a speeding sports car.

The other good reason that our brains are so clever is that they fine-tune themselves. As we saw, whenever neurons talk to "their bosses," their reliability is taken into account. So if two neurons are saying conflicting things to their boss—perhaps one says, "It's a Ferrari," and another says, "It's a Lamborghini"—but the boss knows that the second is more reliable, then the conclusion will be that it's a Lamborghini.

This should sound familiar, because we know how this works. Remember the wineglass with the two straws. Imagine that if the wineglass overflows, the car is a Lamborghini. So the guy on the right is pouring the wine to try to convince you. The guy on the left is drinking the wine, trying to suppress this decision; he thinks it's a Ferrari, not a Lamborghini. But the guy on the right is more reliable—he has a fatter straw—so he pours the wine in fast enough for it to overflow, and you shout, "Lamborghini!"

This is what the brain does. It adjusts the weighting given to stimuli from different neurons, depending on the previous reliability of the signals. So if a particular neuron is very good at spotting Lamborghinis, then its boss (the neuron that receives its signal) will chemically adjust the connection between its dendrite and the neuron's axon to ensure that the signal is stronger. The straws of more reliable sources are made fatter.

By using this method of fine-tuning the signals that travel between neurons in the network, your brain is able to adjust itself so that it can recognize, learn, predict, remember, and indeed do almost everything that the brain does. Clever, isn't it?

So your gray matter works by using vast numbers of neurons in parallel, wired up in a hierarchical network, and changing both the wiring and the strength of signals in the network.

But there are still some mysteries here. For example, I have repeatedly hinted that the brain has centers of specialization—areas that handle only vision, or smell, or language. The neural network in your head is not one big fishing net of connections. It is really a lot of smaller networks that often act independently of each other. It's as though the company *Your Brain, Inc.,* has lots of smaller departments. Why?

ANATOMY OF THE BRAIN

Your brain may resemble a big bowl of pasta (indeed, it may even be the same color and consistency), but this is where the similarity ends. Brains are not all mixed up together, they are divided into distinct regions, each part carefully designed and placed (see plate 4). It is as though your chef has constructed your bowl of pasta by spooning lots of separate heaps into the bowl, taking care not to mix any of them together too much.

We know quite a lot about the regions of the brain. Some years ago (while it was still legal) scientists such as Wilder Penfield (one of the great neurosurgeons and neurologists of all time) electrically stimulated conscious patients' brains. He discovered that he could make limbs twitch, cause sensations of taste, or create hysterical laughter, depending on which regions he applied the electrical current to. Such research is now performed nonintrusively. Neurologists use brain scanners to measure brain activity as patients perform different tasks. Although the scanners are still quite crude (they usually measure only increases in blood flow to neurons when the neurons fire, and have resolutions of around 5 mm square), these methods have been very successful in showing the general brain regions used for different functions.

Today we know that the brain is organized in layers as well as regions. Right in the middle, where the spinal column meets the brain, is an area called the hindbrain. One region here, the medulla, controls breathing, heart and blood vessel activity, swallowing, vomiting, and digestion. Another region, the cerebellum, is a web of neurons dedicated to sensing and regulating movement and posture. Moving further into the brain, we find the midbrain, which is the first port of call for sensory information such as sound and sight and also regulates states of arousal. We then find the forebrain, which is full of regions such as the thalamus, hypothalamus, and telencephalon. The last forms the impressive folded cortex or cerebrum that wraps around most of the other regions. This structure is divided into two hemispheres, with each hemisphere having four lobes. Each lobe also seems to be dedicated to a different function, for example, the frontal lobe (anterior)—the bit under your forehead—handles learning, planning, and other psychological processes. The occipital lobe—the bit at the back of your head—handles vision.

As you can tell, the anatomy of our brains is fiendishly complicated (especially when most of the names of the regions are largely unpronounceable). But why should this be so? Why is the cortex on the outside and the hindbrain in the middle? Why is the vision center at the back of the brain—surely it would make more sense for it to be at the front, closer to the eyes?

EVOLUTIONARY HOUSE EXTENSIONS

Why are there different parts in the brain, and why are they in those locations? The answer is the same I would give if you asked why we have five fingers instead of three or why we have the number of ribs, vertebrae, and teeth that we do. The reason we are the way we are has really more do to with how we evolved than what we do right now.

To illustrate this, let me show you around my home. I live in an apartment in the top-right corner of a big old house. The house is about 200 years old, and during that period it has been changed and extended many times. As far back as I can trace, the house began life as a fairly sturdy nineteenth-century home. Over 150 years ago, an extension was added to the back to provide more space. Later (in around 1860), the whole building was split into two, with the right half used as a police station, and the left half for the constable's house. Still later, the two halves were converted into two shops, and another extension was added to the rear of the first. Even later, more conversions were made, and my apartment was created.

Because of this long history of changes, my apartment has "lots of character," as real estate agents like to put it. Walls seem to be of different thicknesses, windows are different sizes, the floors of half of the rooms are a foot higher than the other half, and the roofs are clearly at different angles and heights. Nevertheless, it makes a very nice, comfortable home—it does everything I want it to do.

Our bodies, and, without any doubt, our brains, have undergone a similar process. Over the course of evolutionary history, as we developed from fishlike creatures to reptiles to small mammals to simple primates to humans, our brains have been enlarged, extended, converted, and reshaped. And just as I can see some of the many extensions and changes that were made to my apartment, neurologists have discovered the many extensions and changes to the brain. It is because our brain is the result of millions of years of alterations that our vision centers have ended up at the back of the brain. It explains why the cortex is on the outside and the brain stem is in the middle, for the brain stem is an ancient part of the early brains of our ancestors, while the cortex is a much later extension. The brain is clearly a wonderfully efficient design, for evolution is the master of making the best of what it has got. But the overall anatomy and layout of our brains are mainly because of the existing structure of our ancestors' brains, not because they necessarily improve the functioning (although they sometimes do this as well).

Indeed, a well-known model describes our brains as triune, or having three parts, consisting of a reptilian brain stem core, an old mammalian brain called the limbic system, and the more recent neocortex. These three elements are layers, each wrapped around the previous one—extensions added by evolution. This layering effect is quite apparent during the growth of an embryo. For example, a forty-five-day-old human embryo has a brain surprisingly similar to the reptilian brain.

By examining life that exists today, we can gain a clear picture of how our brains formed. Let's take a walk down the memory lane of evolution and see where it all came from.

When the first multicellular organisms lived on the Earth, the earliest form of communication among cells was entirely chemical. (We still contain such chemical signaling within us, in the form of hormones.) This communication was improved when new chemicals that allowed the generation of electricity were used, resulting in the first neurons. These allowed simple jellyfish-like creatures to control their movements and react to stimuli. Within us, we still contain such simple neuronal connections. For example, if you burn your finger, you snatch your hand out of harm's way before the signal even reaches your brain. Some special reflexive neurons grab the pain signal and respond, causing the unconscious reaction.

As evolution progressed, networks of neurons were used to control the digestive systems of simple tubular creatures. Indeed, we still have the remnants of this simple brain in our own guts. It's known as the enteric nervous system; some neurologists call this a second brain because it develops separately from our brain in the embryo, being linked later by a cable known as the vagus nerve. Some suggest that this "brain" is responsible for the butterflies in our stomachs when we are about to go onstage, as it mirrors the anxiety of the central nervous system.

Evolution continued its relentless pace, and soon bodies began to be designed in segments, with the main sensory cells clustering at the food intake or head end. The movement of each segment was controlled by clusters of neurons, called ganglia. A central cord connected ganglia to another cluster of neurons in the head. This structure, known as a cerebral ganglion, began to take control of the body and senses.

Moving from invertebrate organisms to vertebrates, the nervous system became progressively more complex. The central cord became encased in bone (our spinal column), and the brain was made increasingly larger to handle the improved senses and new forms of locomotion. When reptiles emerged, new brain centers to handle smell and better vision were created. These structures were then wrapped in newer brain material to handle memory and primary emotions such as love, hate, and fear as mammals formed. Finally, another layer of brain material was added as the structures contained in the neocortex appeared, allowing creatures to plan, understand, and communicate. As the demands of life grew more complex, the cortex expanded and became furrowed to increase the surface area and permit more connections between neurons.

This is the history of our minds. Like the uneven and unplanned changes made to my house over the years, our brains have been repeatedly extended and reshaped. Each new feature has been added to what was already there, resulting in the complex anatomy our brains display today.

And at some point during the evolution of brains, they became so complex that they began to be aware of themselves. They developed consciousness.

CONSCIOUSNESS

Consider yourself. You are a living creature aware of your own existence. You know when you are happy or sad. You know when you've made a mistake, or when you have succeeded in solving a hard problem. You know you are an entity that is separate from your surroundings. You have your own thoughts and your own opinions. Indeed, when you speak, you say, "*I* think that . . ." You don't say, "The human over here thinks that . . ."

But what are you? What creates that special uniqueness that is your individual self? What creates your self-awareness, your consciousness? Are you a brain? Or a mind? Or a soul?

These are questions that have puzzled scientists and philosophers for centuries, but we are now beginning to see the first glimmers of answers. The answers are being provided by examining people who show deficiencies in their awareness. These deficiencies are caused by faults in the brain.

SEEING IS NOT BELIEVING

He petal tribes our index (backward back the lad) handle sensory information. They inform very special constructing "old model" brings see, following now that sings list stain positions found. Sometimes, transferers likes lose nationality hand petal tribe. Though they fill rings found them correctly well, this rinses them able sink out things their hand ride. Spite ring seriously able objects sir deft, they able incorporate these objects to their old model. Thing left lists pour them. You right imagine, this likes weeding hairy lard lunch patients. Bother parts their drains don't they multiply card sort words went words Tom light half that they.

Is the previous paragraph a printing mistake? No. This is what it is like to read if you suffer from a certain brain disorder. Here is what I actually wrote:

The parietal lobes of our cortex (toward the back of the head) handle sensory information. They perform the very special job of constructing a "world model" of the things we see, allowing us to know that things exist in certain positions around us. Sometimes, sufferers of strokes can lose the functionality of the right-hand parietal lobe. Although they can still see things around them perfectly well, this causes them to be unable to think about things on their left-hand side. Despite being consciously able to see objects on their left, they are unable to incorporate those objects into their world model. So nothing on the left exists for them. As you might imagine, this makes reading very hard for such patients. But the other parts of their brains don't let on—they simply discard short words and invent words from the right half that they get to see.

A little clearer now? Well, give thanks to your parietal lobes, for it is they that are doing much of the automatic, unconscious processing as you read this.

We care about such disorders because they tell us about the conscious and unconscious processes in our brains. We discover just how many things that we do are performed unconsciously, without our being aware of what is going on. We also discover how important the unconscious activities are in allowing us to build our conscious awareness of our surroundings. So for patients with a malfunctioning parietal lobe, their entire awareness of the world becomes distorted. Ask them to draw a flower, and they will draw only half a flower. The unconscious activities of our brain enable and expand our conscious activities.

We've just seen what happens when a parietal lobe of the brain does not function properly. But as we read earlier, the signals from our eyes also go to a second region of the brain: the visual center or occipital lobes. This region of the brain gives us our conscious view of the world: it allows us to be aware of what we see. And some victims of strokes lose the functionality of an occipital lobe.

BLIND SIGHT

A patient stares at a computer screen. A little square appears and moves up. The patient says, "Up, not aware." A little square appears and moves down. "Down, not aware." The square appears and moves to the right. "Right, not aware," he says.

This patient is unable consciously to see anything on his right. A car accident damaged his left occipital lobe, and he no longer has the part of his brain necessary to process all the information from his eyes. He is, to all intents and purposes, blind on that side. And yet he is able to guess correctly the movement of objects he is not aware of seeing.

But he is not using ESP, he is using his parietal lobes. Although he is oblivious to what is going on, his parietal lobes are performing unconscious processing of the signals from his eyes. Although he sees nothing, he still unconsciously knows about things like movement in front of him.

This condition has been named *blind sight* for obvious reasons. It shows, once again, how our consciousness is affected by damage to the brain. Instead of being aware of seeing objects but not being able to grasp that they exist, these people are unaware that they are seeing anything but know unconsciously that something is there and moving.

Before we work out how this helps to explain how consciousness happens, let's look at one final condition that we began the chapter with.

TICKLE TESTS

Sarah Blakemore and Chris Frith didn't just tickle patients for fun, they were testing a theory of conscious behavior put forward by Chris.

Professor Chris Frith believes that we move through the world by generating two models within our brains. First, we create an "inverse model," where we work out which commands to send to our muscles to produce the movement. Second, we generate a "forward model," where we work out where our limbs will be and what the sensory input might be given those commands. So every time we consciously move, we work out how to do it and what will happen once we've moved. We can then do comparisons with what actually happens and what we thought would happen, to allow us to fine-tune our future movement.

The tickle tests were performed to investigate these ideas. When you tickle yourself, you predict both the movement and the feeling of your hands, which removes the surprise, making the activity rather less funny. If you tickle yourself using a robot that responds instantly to your commands, you also predict the movement and response, and so you don't laugh. But, as we saw earlier, if the movement of the robot is just slightly delayed, your prediction is thrown off, and the tickling becomes funny again.

But some people with schizophrenia are able to tickle themselves all the time. So it may be that these schizophrenic patients have a minor malfunction in their forward model. Although they can work out how to move consciously in a new way, they sometimes do not correctly predict what the result of the movement will be. This is why they feel as though their movement is being produced by an external force: the movement of their own limbs surprises them. It also explains why they can sometimes tickle themselves without difficulty.

Again, we have an example of unconscious processes in the brain (making models and predictions) that can affect our consciousness. But enough of tickle tests and blind sight. It is time to tackle the question head on. What is consciousness?

CROWDS OF SELF

The soccer crowd is silent and slightly restless. To relieve the boredom, a few people begin the "wave," standing up with their hands above their heads, then sitting, in turn. The wave spreads, with more and more people joining in, and before long, a giant wave is rushing around the stadium. Suddenly there is some action on the field; the crowd gasps as one as a goal is narrowly missed. The play dies down again. Triggered by the near miss, some people begin singing. The song becomes louder and louder as more and more of the audience joins in. Soon the deafening sound of tens of thousands of voices all singing the same song can be heard. Then thousands of simultaneous groans as a player is fouled.

We are used to this kind of crowd activity—thousands of separate people unified by watching a game, their focus shifting from the wave to singing, to watching, to celebrating. And yet everyone is an individual, doing his or her own thing. There is no one in control, ordering, "Do the wave!" or, "Everybody sing!" or, "One-two-three, groan!" The crowd concentrates on new things dynamically as people start a new song or see something dramatic happen.

This is what you are made from. Your mind, your consciousness is the "crowd activity" of all of the billions of neurons organized in their different regions in your brain. There is no single spot in the brain where your self lives. Your consciousness is made from all of your brain.

Like the soccer crowd, separate neurons do thousands of things that do not directly impinge on consciousness. The people in the crowd are breathing; they also may be eating, on their way to or from the rest room, on the phone, or doing many other "unconscious activities"—unconscious, that is, to the crowd as a whole. Your neurons are doing substantially more activities than these, such as the movement of your eyes and translation of these words as you read. These are unconscious activities.

Also like the soccer crowd, your awareness, your focus of attention, is constantly changing. Just as a few people singing can cause the entire crowd to "think of singing," a few neurons firing at the right time will cause a spread of neuronal activation and you will think of something. When a crowd sees something dramatic, its focus switches instantly—even when many of the audience missed the event. Similarly, when the group of neurons in your visual center reports something unusual, the focus of most of the neurons in your brain switches instantly.

And what of self-awareness? A larger crowd is certainly aware of itself: if some people are acting inappropriately, others will not be impressed. Likewise, if you find yourself saying something inappropriate, you are immediately aware that you made a mistake. Your embarrassment is caused by one part of your crowded consciousness chastising the part that just made the mistake.

As a migraine sufferer, I know firsthand what it feels like when you see flashing lights at the onset of a classic migraine. One part of my brain is realizing that another part is not working properly. In a similar way, if people on the other side of the stadium are being hurt by overcrowding, the people on this side will react with alarm—another example of self-awareness.

So this is you: a crowd with ever-changing ideas, both conscious and unconscious, that dynamically focuses on new things, be they ideas, its surroundings, or itself, without any single person in charge. Your self is created by the crowd of neurons, connections, and brain regions inside your skull.*

•　　•　　•

---

*These kinds of views of consciousness are becoming increasingly accepted by scientists of many fields. Daniel Dennett does a good job of explaining his version in the book *Consciousness Explained*. However, some scientists still prefer to avoid using the "c" word altogether. When I spoke to Lewis Wolpert on this subject, he said, "My rule about consciousness is that it's a word that can only be used in my presence with a permit—which I give out very, very rarely." He continued in a whisper, "I think all the stuff about consciousness is bullshit!"

And this completes our journey through our biological brains. We've seen what neurons are and what they do. We've looked at how the neurons are wired together, and how a neural network can learn. We've discovered the anatomy of the brain and why such regions exist. We've seen how all of these things working together create our consciousness. Having made this journey, we must now move on to another kind of brain, a brain that is also made from neurons but is not gray or mushy—digital brains that think inside the digital universes of our computers.

BRAINS IN COMPUTERS

The linguistic philosopher John Searle was not terribly impressed with our earliest attempts at machine intelligence. He was so unimpressed that he created a thought experiment, the Chinese Room, from which some computer scientists have been trying to escape ever since.

The thought experiment went like this. John imagined himself sitting in a room. He would be passed Chinese symbols through a window. He would look up the symbols in a rule book, which would tell him to find another symbol in one of the giant filing cabinets in the room. He would then pass that symbol through the window. At no point would he understand any of the symbols.

But the people on the outside of the room had a very different view. They were passing their questions in Chinese through the window, and valid answers were being returned to them. They were convinced that the room contained someone with intelligence who understood the meaning of the questions.

John described this thought experiment to show that computers might appear to act intelligently, but they would never understand what they were doing. It would always be symbol processing, never comprehension, so computers could never truly become intelligent or aware.

And he was right—or at least, partially right. At the time, in 1980, there were computer programs called expert systems that worked exactly like this. They responded to input symbols by finding and returning the appropriate output symbol.* So although they would be

---

*We will return to ideas of symbol processing in the next chapter.

able to answer the question "What is the current time?" with the right answer, they would never know what the concept of time was.

But today, "machine intelligence" is no longer a misnomer. We now use software to create digital universes in which digital entities exist. And when we want our computers to behave in intelligent ways, we allow digital neurons to exist. These neurons work in similar ways to the neurons in your head, and they are wired up into neural networks, as they are in your head. So we really do have digital brains (albeit simple ones) that exist inside computers. And these brains really do learn, predict, identify, control, remember, and do a hundred other things.

Like the separate neurons in our brains, digital neurons act predictably, firing when the combination of their input signals passes a threshold. So although individual neurons, both biological and digital, have no idea what is going on (like John in his Chinese Room), the crowd of neurons working together is capable of understanding. As confirmed during my discussion with the internationally renowned computational neuroscientist Geoffrey Hinton, neural networks within our computers may be inspired by nature, but they are not fakes or simulacra. A neural network within the digital universe of our computers works by following the same underlying processes of biological brains and so deserves the title of digital brain. Indeed, when I asked Geoff Hinton whether we will one day have digital brains of complexity and abilities similar to biological brains, his response was an unequivocal "Yes. I'm sure we will."

Now that we've cleared that up, let's look a little more closely at a digital brain. Just what do these nonbiological intelligences look like? How do digital neurons work?

DIGITAL NEURONS

There may be a hundred types of neuron in your head, but there are many more different types of digital neuron. Unlike nature, the different digital types are usually not used together in the same neural network.

As with the biological neurons, I have neither the space nor the perseverance to describe every different type of digital neuron, so we will not look at the exceedingly complex digital neurons (often described in the form of electronic circuits) used by some computational neuroscientists to model biological brains. Nor will we look at the exotic varieties used in computer science for the thousands of different neural network applications around the world. Instead, we will look at the most common features found in most digital neurons.

If you recall, our average, run-of-the-mill, standard kind of biological neuron had three important parts: the cell body, the axon, and the dendrites. A digital neuron comprises much the same parts, only made from software. The "cell body" is often called a processing element, or simply a neuron. This has a single output and multiple inputs (corresponding to the axon and dendrites, respectively).

Most (but not all) digital neurons do not send or receive pulses through their outputs and inputs. Instead, they send out steady signals— for example, a 0 or 1—and they receive much the same from sensors or other neurons. Sending a continuous signal is intended to be equivalent to the series of pulses that a biological neuron would emit, and indeed, this approximation is acceptable for many applications.

Apart from this simplification, a digital neuron works in much the same way as the biological neuron. Remember the wineglass with two straws? When the total wine entering the glass exceeded the capacity and spilt on you, you jumped up and shouted, like a neuron firing a pulse when its inputs exceeded a threshold. A digital neuron works in exactly this way. Given a series of inputs, the neuron will calculate the weighted sum and might output a 1 if the value is above a threshold or 0 if the value is below the threshold.

You'll notice I said "weighted sum" there. If you recall the wineglass with its two straws again, the straws could be of different thicknesses. This corresponded to our neuron paying a little more attention to the signal from one neuron compared to another. To achieve the same straw- thickening effect with digital neurons, every input has a weighting value. So although two inputs may be sending a 1 signal to the digital neuron, if the left input has a weight of 0.5 and the right input has a weight of 1.0, then the sum of weighted inputs will be $(1 \times 0.5) + (1 \times 1.0)$, which is 1.5. In other words, our left input has a straw half the thickness of the right input, so the right input has twice the effect. To continue one step further, if the threshold was 0.75, then the neuron would output 1 in this example, for the sum of its weighted inputs is

1.5, which is higher than the threshold. And if we wanted one of the inputs to inhibit the neuron rather than excite it, we would simply use a negative weight.

In fact, digital neurons are usually more complicated. Instead of outputting a 1 or 0 depending on whether the weighted sum is above the threshold, many neurons calculate the weighted sum of the inputs, subtract the threshold, and then output this value directly.* By doing this, the digital neuron behaves a little more like a biological neuron, where a high output value corresponds to a fast series of pulses and a low output value corresponds to a slow series of pulses.

It turns out that even this kind of neuron (known as the perceptron) is slightly limited. Because its output corresponds to its input in a linear fashion, it can learn only very basic, linear functions. This means that any change in the inputs corresponds to a similar change in the output. So a small change in one input would always cause a corresponding small change in the output. The output is proportionate to the input, making these perceptrons rather boring and ineffective.

To overcome this, a typical digital neuron has one extra trick. Once it has calculated its sum of weighted inputs and subtracted the threshold, it transforms the value. This is performed by a transfer or activation function, typically using some form of sigmoid curve. If you like math, a common function is $y = 1/(1 + e^{-x})$. If, like me, this does not immediately say very much to you, don't worry—it's just a way of making the neuron output a value between 0 and 1 that depends on the inputs in a nonlinear way. Which means that the neuron is now nonlinear—it can learn much more complicated things.

So this is our standard digital neuron. It lives inside our computers, accepting numerous inputs, each a value between 0 and 1. It weights and sums those inputs, and subtracts the threshold value. It then transforms this total using a nonlinear sigmoid function and outputs the result, which will be a value between 0 and 1.

And the values to the inputs are sent from other neurons that are arranged in a neural network.

---

*But it outputs the value only if it lies between 1 and 0. If it is more than 1, it outputs 1; if less than 0, it outputs 0.

NEURAL NETWORKS

The neural networks behind your eyes were grown following instructions that took evolution millions of years to perfect. We are still trying to work out how to connect networks of digital neurons to make them perform different tasks. But in the past fifty years of using digital neurons, we have learned a thing or two about which networks seem to work.

There are two main categories of connections used for digital neural networks: *feedforward* and *recurrent*. Feedforward networks have neurons in rows or layers. Rather like some of the networks we have linked to our retinas, they connect neurons in the forward direction only. The first layer of neurons will be connected to the next layer, those will be connected to the next layer, and so on. Neurons in one layer are never connected to the neurons in previous layers, hence the name *feedforward*. On the other hand, recurrent (or feedback) networks allow connections between any neurons. Although still often arranged in layers, these networks permit later neurons to feed their outputs back into the inputs of previous neurons.

Many researchers prefer feedforward networks, as we now have a considerable amount of theory and mathematics to predict and explain their behavior. However, recurrent networks are capable of more complex activities and so tend to be favored by those who aren't so concerned about the lack of theory of such networks, but do want dynamic control for their robots, for example. For now, let's just worry about feedforward networks and how they might solve a problem.

Imagine you own a factory that makes plates. Your plates are automatically made in molds, sent to kilns where they are fired, have patterns printed on them, are fired again, and are boxed—all by machines. The trouble is that the machines are not perfect. One in every twenty plates will have cracks, stains, or misplaced patterns on it. Because of the speed of production, you need to employ ten people full-time to examine every plate as it comes off the production line. And because it is such a dull job, they make mistakes, and poor-quality plates still get shipped to customers. What you need is an automatic system that can look at the plates and judge whether they are acceptable or not. So you set up a camera and tell a computer to compare the images the camera sees with a photo of a perfect plate. Your computer rejects every plate. The problem is that the camera sees a slightly different plate every time: they are lit differently, they are upside down, they are ever so slightly different colors. The computer is not intelligent enough to understand the difference between a bad plate and a plate seen under different lighting conditions.

So you decide to make the computer more intelligent. You give it a feedforward neural network. Each pixel from the camera image is a number signifying the color. You feed these numbers into a layer of neurons.* The output of those neurons feeds into another layer, which feeds into another layer, and so on. Eventually, the outputs of a layer feed into a single neuron, which has one output. When the output of this neuron is high, you decide that the plate will be considered acceptable. When low, the plate will be rejected.

You now have a digital brain that can examine your plates and say "okay" or "reject." How does it know which plates are good and which are bad? You have to teach it. You provide the neural net with a collection of pictures showing good and bad plates. As it looks at each one, you tell it, "This one is good" or "This one should be rejected." And it *learns.*

After a while, it is getting the answers right without your needing to tell it. You now let your trained neural network look at images of plates it has never seen before, and it correctly tells you which are okay and which should be rejected. Your computer with its digital brain now takes over the job of quality control in your factory. It never gets bored, it rarely makes a mistake, and productivity increases—allowing you to employ the ten people who were on quality control for sales instead. Don't you love a happy ending?

But hang on. Just how did that neural network learn?

If you cast your mind back, we discovered earlier that a biological

---

*It would be more sensible to do some preprocessing of the images first—for example, make sure that the plates were all in the center of the images and the same size.

neural network is a little like a company. All employees report to their immediate bosses, who report to their bosses, and so on. You should also remember that each boss remembers those employees who gave the best suggestions, ensuring that they would be listened to a little more next time. The biological neural network learned by chemically adjusting the strength of the connections between its neurons, so that the most appropriate signals for a given situation have the greatest effect.

Not surprisingly, this is also how our digital neural networks learn. In the example, every time we presented the feedforward neural net with a picture of a plate and said, "This one is good," the network examined its output value. If the output was not high (remember that high values correspond to good plates), then the network adjusted the input weights of each neuron, until the output was high. And every time we presented a picture of a bad plate and said, "This one should be rejected," the network examined its output value. If the value was not low, it adjusted the input weights of the neurons until it became low. Over time, the weights of the neural network settle onto specific values—some high, some low, some medium—different for every neuron. The network now contains knowledge about your plates embedded in the different weighting values, just as the patterns of wear in your carpet contain information about where you normally walk. So when you present the trained network with images it has never seen before, it simply uses its learned knowledge to output the right value.

"And exactly how does the network adjust its weights?" I hear you ask. A very good question. Unfortunately, we currently have no idea how our brains manage to do this, even though we know it does happen. So computer scientists and neuroscientists have had to make up their own methods, which they call learning rules. Some researchers have used evolutionary algorithms to evolve the weights. Whichever method is used, they all work along similar principles: the input weights of neurons are iteratively adjusted in small steps, while checking to see whether the output is improved or not. It's rather like having a few hundred knobs to tune in your television set—you keep adjusting different knobs by very small amounts until you're happy with the picture.

The learning rules, types of connections, types of neurons, and types of activation function all determine the type (and name) of the neural network. There are now hundreds of different kinds of these digital brains in use, all designed for different applications.

THOUGHT FOR BUSINESS

Out of all of the biologically inspired computation methods described in this book, neural networks have been around the longest and used the most. There are quite a few companies in various parts of the world that do nothing but create digital brains to solve other companies' problems. The number of applications is too great to count; neural networks are now regarded as just another tool to be used for particularly challenging problems.

Using a neural network for quality control of plates is actually a very common task for a neural network. Hundreds of factories use systems just like these for checking their products. From electronic circuits to clothing, digital brains are watching over our products right now. Indeed, some years ago, Science Applications International Corporation (SAIC) added a neural network into baggage checking at New York's JFK International Airport. Called SNOOPE, it was trained to detect the characteristic gamma ray emissions of explosive materials. Unlike the "stupid" machines before it, SNOOPE could tell the difference between the nitrogen in a bomb and the nitrogen in cheese, for example. When it found anything suspicious, the bags were automatically rerouted to a human inspector.

Neural networks have been used for financial applications for even longer than this. Credit cards are checked for fraud, insurance claims are assessed, and customer profiles are found. One of the earliest neural network companies, Nestor, had a product that appraised mortgage applications. Its neural network was trained on several thousand applications—some which had been accepted and some rejected. The network learned to predict which applications were too risky to accept, and in tests where its performance was compared with human underwriters, it was found to be more consistent. Today many companies specialize in providing such intelligent software for the financial sector.

There seems to be no end to the problems that neural nets have been applied to. Forklift truck controllers, sonar mine detectors, handwriting and speech recognizers, tumor identifiers. All performed by digital brains. And we shouldn't forget the neural networks used by neuroscientists to model and understand the workings of our own brains. But we must move on, so I'll mention only one more application before we leave the world of neural networks. This is perhaps the oldest and most successful of all, developed in the 1950s by Bernard Widrow, winner of the IEEE Neural Networks Pioneer Medal in 1991. Benard was a pioneer in the field, and developed the first adaptive noise-canceling system for the telecommunication industry. His neural-inspired adaptive filter cleaned up the echoes on telephone lines and reduced transmission errors for modems. In the days before digital communication, this was essential to allow proper communication, so his adaptive filter was incorporated into telecommunication systems worldwide and used for over thirty years. Digital brains have been listening to your conversations and making them easier to hear all that time.

This is the exceedingly successful digital brain. Used for hundreds of problems, it is but a collection of neurons wired up in a network that lives within the digital universe of computers. But this is not the only type of digital brain.

ANATOMY OF ROBOT BRAINS

We saw earlier that our brains are not one giant neural network. They have distinct anatomies, with regions of specialization. The gray matter in our heads has neural networks for vision, speech, movement, hearing, memory, and everything else that we do. Many of these neural networks perform their unconscious activities largely independent of each other, like separate computers working in parallel.

Marvin Minsky at MIT takes this idea to an extreme. He believes that we should regard the biological brain as a very complex machine built from a series of smaller, largely independent computers or "agencies." (He also believes that consciousness is largely illusionary, but we will return to this later.) He calls his theory the society of the mind, where the separate brain regions act like members of a society, cooperating through separate activities to help the whole.

Perhaps influenced by such theories and also through many years of research on robot control, Rodney Brooks, also at MIT, has developed a new style of digital brain. These brains are usually physical devices—electronic circuits and computers that control a range of real robots, some humanoid in appearance. And all of these brains follow a new paradigm, the *subsumption architecture*.

Just as our brains have anatomies, with areas of specialization, Rod's robot brains have anatomies with separate circuits and computers performing independent tasks.* As we saw previously, the anatomy of our brains is mainly the result of our own unique evolutionary history. Similarly, Rod and his team build their robot brains by an incremental process akin to evolution.†

Instead of having a single neural network or a traditional computer-based planner that processes inputs, develops world models, and produces a single output, the subsumption architecture enables a wholly new form of robot brain to be developed. These brains work in parallel, multiple processors examining the sensors and suggesting actions at the same time.

To see how this works, let's look at the first robot developed using the subsumption architecture, "Allen." Looking like a motorized trash can, Allen was capable of moving toward distant goals while avoiding obstacles. And yet Allen had no centralized controller to form a plan and

---

*Rodney Brooks takes pains to point out that the anatomies of his robot brains are not directly inspired by the specific anatomy of our own brains. Since we do not really understand the function of each region and how they work together, he prefers to let the robots develop their own unique anatomies as he and his team slowly add new behavior-generating modules. He says, "Skepticism should be applied" to the brain functional-decomposition approaches.

†Akin to, but not the same as. The researchers may provide variation and selection, but they do not use populations of reproducing brains that have children that inherit parents' features. Their process of brain development is an incremental additive process, not true evolution.

deliberately navigate around objects. Instead, it had a digital brain that consisted of three simple layers of control. The first layer used Allen's sonar. If anything got too close to the robot, Allen would move away to a safe distance. In addition, if Allen was moving forward and the sonar "saw" something ahead, the robot would halt. The second layer subsumed the first layer—every ten seconds, a desire to wander about randomly was generated. Although the obstacle avoidance of the first layer still activated if Allen got too close to objects, the direction of movement was now dictated by the second layer. Finally, the third layer looked for distant places and made the robot move toward them. Again, this subsumed the other layers, specifying the direction of movement (unless obstacles got in the way, when the other layers would influence and change the direction).

So the movement of Allen was created by three separate processes within its brain: "don't bump into anything," "explore," and "head toward a target." Together they enabled the robot to discover a way past objects and perform clever route finding to reach destinations.

The subsumption architecture, with each layer of control subsuming another below, is designed to correspond to one idea of how the separate regions in our brain work together. For example, we have largely automatic layers in our brain to take care of walking, but these can be overridden by a desire to jump in a puddle, which can be overridden at the last second when we realize how deep the water is. So Brooks's robot brains contain processes, each working separately and each capable of overriding or influencing the effect of the others. Interestingly, these processes are not always neural networks. Rod believes that the overall architecture or anatomy is more important than the construction of the individual processes.

The parallelism of many simple processes all doing their own thing at the same time makes these robot brains surprisingly fast and adaptable. To show this, the MIT Artificial Intelligence Laboratory has created progressively more complex robots since the days of Allen. Tom and Jerry (which looked like two toy cars) were able to chase objects without ever knowing that there were such things as objects. Herbert (which had a brain made from twenty-four simple processors) was able to trundle into people's offices and steal their soda cans. Genghis was a six-legged robot with separate controllers for each leg, which managed to walk over rough terrain without difficulty. But the robot that has received perhaps the most attention is Cog, a life-size humanoid robot that has been slowly built over the past fifteen years (see plate 5). Cog now has two arms, an articulated head with cameras, which it moves with the same proficiency as we do when we move our eyes, and a huge number of processors arranged in a subsumption architecture. Cog has reflexes, can nod its head, and is capable of following objects with its eyes and learning how to reach for them. It can also recognize faces, and the team are now attempting to add behaviors for social interaction. As more behaviors are added to the robot, the researchers in the "Cog shop" liken Cog's development to that of a growing infant.

While the MIT researchers developing their robot brains may take the role of evolution, others use real evolution to create digital brains (see plate 6). And some believe this may be the route to one of the most controversial possibilities to arise from our technology: conscious computers.

EVOLVING TO CONSCIOUSNESS?

There has long been a small but vigorous fusion of research in evolutionary computation and neural computation. Earlier I mentioned that evolution has been used to adjust the weights of neural networks during learning. More recently, evolution has been used to design entire networks, specifying how many neurons should be used and how they should be connected. Such applications are most common in the field of robotics, with researchers evolving the sensory systems and brains of physical robots.

EVOLVING BRAINS

A nice example of this is the work of Inman Harvey, Phil Husbands, and Dave Cliff, then all at Sussex University. They enabled evolution to generate a recurrent neural network and the "visual fields" (areas receptive to light) of the robot's eyes. Their "gantry robot"—a robot suspended from above—quickly developed a brain and eyes capable of distinguishing between a triangle and a rectangle. This was seen to be particularly impressive when the researchers discovered that the robot only used two visual fields, as though it had compound eyes consisting of only two elements.

Another memorable piece of research in this area was performed by evolutionary artist Karl Sims.* Karl used some of the most powerful computers of the time to create a digital universe with many of the laws of our universe. Gravity, friction, the law of moments, and inertia, as well as light, water, and land, were all re-created. Within this environment, Karl allowed "virtual creatures" to evolve. Starting with very little brains or bodies, evolution developed both, resulting in turtle-like creatures that could swim by moving their flippers and snake-like creatures that spiraled through the water. On land, there were creatures that ran along at high speed, hopped, or tumbled in bizarre ways. As an experiment, Karl took the swimming snakes and re-evolved them for land. They quickly developed a movement exactly like sidewinder snakes in nature. Since this remarkable research, Jeff Ventrella, a specialist in computer graphics and artificial life, has duplicated these results, this time using modern desktop computers. His program is freely available from his Web site and makes a fascinating distraction as new swimming creatures evolve before your eyes. Other researchers, such as Jordan Pollack and his team at Brandeis University, are now attempting to evolve the brains and bodies of physical robots, moving from the digital universe into our own.

The ability of evolution to create digital neural networks and digital body shapes that look and behave spookily like creatures in nature is amazing. Equally amazing is the complexity of the evolved digital brains. Sims delighted his audiences by telling them that his virtual creatures had neural networks so complicated that drawing just one of them would require a room-sized sheet of paper, and understanding them would be next to impossible.

And yet we must look at such work in context. Few neural networks in our computers use as many as 100 neurons. Many of Karl's creatures

---

*I was in the audience some years ago when he first presented his results, and this was the first and last time I have heard a standing ovation at a scientific conference. Since then biologists such as Richard Dawkins and Lewis Wolpert have also expressed their enthusiasm. When I described this work to Lewis, his reaction was "I love it!"

did not exceed this number. But the average slug has 350 neurons, so our typical digital brains are less intelligent than a half-wit slug. Admittedly, some of the more complex digital brains, such as Cog's, might be as clever as a small crowd of snails, but we're still not exactly talking about digital masterminds here. Remember that our brains have something like 100 billion neurons. Clearly, if we want to achieve that highly controversial aim of conscious computers, the level of complexity of digital brains must be increased by a factor of a million or so.

These problems have been noted by researchers, and some have dedicated their careers to creating such vast neural networks. Perhaps the most difficult-to-miss researcher in this area is Hugo de Garis, formerly head of the "Brain-Builder Group" at Starlab, in Brussels. Hugo is a man with a vision. He wishes to develop the first digital brain with a billion neurons and have this brain control a life-sized robot kitten.* The ideas are genuinely clever: he and his team are using methods from evolvable hardware (mentioned in the previous chapter) to both evolve and grow neural networks in real electronic circuitry. Because the reconfigurable chips can be electrically rewired to function as different circuits so quickly, the chips can behave as a huge number of different neural networks, quickly reconfiguring themselves to act as new networks within the same digital brain millions of times a second.

So what's the problem? Well, Hugo is a real visionary, but it seems that his ideas may be rather beyond our current technology. His group has recently revised its target of a neural network with a billion neurons to a network of 75 million neurons. The research community waits with bated breath to see what happens next.

Making complicated brains is hard. Making them work is equally hard. So what of computer consciousness? Will it ever happen?

STRANGE CONSCIOUSNESS

We are now standing on very well trodden ground. From eminent philosophers and computer scientists to somewhat deranged fanatics, it seems that everyone has a view about this. Our friend Hugo, for exam-

---

*Why a kitten? I don't know. Hugo says it's "purely to show off the capacities of the brain," but I suspect probably because they're cute.

ple, says in some of his recent writing, and I quote directly from his personal Web page: "I have 2 major life goals. a) To build artificial brains and b) To raise the alarm on a possible gigadeath artilect war." According to Hugo, the latter is a risk that "artilects," or artificial intellects, will not only surpass human intellects; they will decide to go to war with humans, leading to "the annihilation of the human population."* And believe it or not, there are even more extreme or unlikely views on this subject. For example, some claim to have developed a new form of electronic circuit that enables "intelligences from other realms" to endow robots with consciousness. Whether this means the souls of people no longer with us or alien intelligences, I couldn't tell, nor did I think it worth the effort to find out. Others try to explain supposed phenomena such as "thought transference" and "collective consciousness" by referring to Einstein's General Theory of Relativity.

But there are some more carefully thought out ideas on computer consciousness surrounding us as we stand here. We've already heard John Searle's objection—the Chinese Room argument—and why it is a little out of date. Alternatively, the physicist Roger Penrose of Oxford University feels that because we make intuitive leaps that "mere machines" cannot, there must be a nonalgorithmic basis to consciousness. Being a quantum physicist, he believes, perhaps not too surprisingly, that the strange laws of quantum mechanics help create our consciousness. Most suspect that his search for an answer is getting a little far from the truth.

Marvin Minsky of MIT believes almost the reverse of Penrose. He thinks that our brains are machines and that most of our impression of being conscious is an illusion. Unfortunately, Marvin takes a rather more limited view of what the word *consciousness* means. He feels that to be conscious is to know how you are thinking something—a detailed internal examination of yourself as you think. Since none of us pays much attention to how or why we think certain things, Marvin believes

---

*Ironically, these visions do not stop Hugo from trying to create as many "artilects" as possible. When I last saw him, his latest prophecies included a plan to create thousands of superintelligent robots that would sow genetically modified plants in a desert in order to transform the land. No, I didn't understand why you couldn't just use a tractor. . . .

we are not terribly conscious most of the time. Indeed, he says, "We humans do not possess much consciousness. That is, we have very little natural ability to sense what happens within and outside ourselves." And because our computers are able to examine their own internal "thought processes," he even states that "some machines are already potentially more conscious than are people."

But Steven Pinker, a cognitive neuroscientist at MIT, points out that "it's a very different sense of the word that captivates people. That sense is sentience: pure being, subjective experience, raw feels, first-person present tense, 'what it is like' to see red or feel pain or taste salt." And we know how this kind of consciousness arises, for we saw it earlier in the form of a soccer crowd reacting to itself and its surroundings. Our consciousness is the result of a crowd of neurons, connections, and brain regions competing for attention, which causes new groups or "agencies" to form and focus on new things dynamically, without any central control. It all happens inside the gray matter of our brains, following the processes we have explored in this chapter. And when those same processes are used within the digital universes of our computers, given appropriate numbers of neurons, connections, and anatomy, there is no good reason why a digital brain could not attain consciousness every bit as valid as our own.

It's certainly possible, but will it happen? Will we ever be able to construct or evolve digital brains that work this well? I don't see why not. But I suspect it may take a while.

SUMMARY

Understanding brains is a ticklish subject. But science is beginning to shed light on the mysteries within our heads. We now know that our brains think and learn using neurons—neurons that act just a little bit like a wineglass with straws in it. These neurons are cleverly wired together into networks, each of which can then learn a little like a company with many levels of employees and bosses. The brain is also made from many different regions, and those regions were slowly added by evolution like the extensions to a building. Finally, consciousness is made from many processes, some unconscious, some not, that emerge like the crowd activity at a soccer game.

Digital brains are much less complicated, and we know—mostly—how they work. Instead of Chinese Rooms, digital brains are usually found within the digital universes of our computers. They are also made up from neurons, accepting signals that they weight, sum, adjust with a threshold, and transform with an activation function. They are also wired into networks—some feedforward, some recurrent—and they also learn. For example, a neural network might be used for quality control in a plate factory, learning by adjusting the input weights of every neuron to make the output match training examples. These neural networks work very well, being used for huge numbers of applications such as bomb detectors, mortgage evaluators, and telephone noise filters. Not only that, but the subsumption architecture has resulted in robot brains with anatomies—specialized regions or processes that perform parallel independent tasks. And finally, the thorny question of computer consciousness. It seems likely that a digital brain within a digital universe may one day—if we get everything just right—be conscious.

So that's brains for you: neurons, networks, and regions leading to thought. But there are other kinds of brains. Brains that do not have neurons and that think in strange ways. Brains made out of insects.

5

INSECTS

| Darkness | Hard, solid walls | Enclosing, constricting | Not enough space | Must break free | Pushing, stretching | Walls splitting | Ah! The smell! | The smell of home | And | Sweetness, food. Hunger|

With a final wriggle, the princess was born. Her white oval body rejoiced in the freedom from her egg, now too small to contain her growing form. Still barely aware of her surroundings, she was picked up by something and quickly moved, through dark corridors and past endless streams of other scurrying forms. She was then deposited next to other things that wriggled nearby. The smell of sweetness was now overpowering, intoxicating.

Without warning, mandibles were pushed against hers, and she tasted sweetness flowing into her mouth. It was over far too quickly. More! She needed more! But although her feeders returned often, her appetite seemed only to increase every time she was fed.

Soon her oval body far dwarfed the size of her feeders, which now needed to climb over her to reach her mouth parts. She was also aware of new smells. The odors seemed to be special, informing her of her heritage, her destiny. The scents were harbingers of her future. She would not be a worker like those that fed her. She would be royalty.

As she fed and grew and fed and grew, she slowly became more satisfied. The hunger at last abating, she became still, her skin hardening around her now-impressive oval larval form. Within the hardened shell, her body began to change, following the instructions in her genes triggered by the scents she had experienced. Like a pearl hidden within the shell of an oyster, her lustrous new form was built in secret blackness, invisible to the outside world.

| Darkness | Hard, solid walls | Enclosing, constricting | Not enough space | Must break free | Pushing, stretching | Walls splitting | Ah! The smell! | The smell of home | And sweetness, food | Hunger|

With a new body that recalled nothing of its previous brief life, the princess crawled from her pupal shell. Her new head moved around for the first time, trying to untangle its antennae as they unfurled. Her six new legs allowed her to stand proud as her glorious wings—the mark of her royal status—gradually uncrumped, expanded, and hardened behind her. In the darkness around her, she became aware of countless others like her, approaching her and touching antennae with her.

Although the temperature in the chamber did not seem to vary where she was, she could sometimes detect that the workers had come from somewhere warmer. On these occasions, activity would increase, with all the workers and princesses jostling around and moving incessantly. As more princesses emerged, space became less and less, and she could now no longer move without finding others touching her. The sense of urgency increased, and she found herself moving more and shaking her wings to exercise them in the cramped space.

At last the day came when the air tasted like water, and the workers were hot, frantic, and excited. There were thrilling new odors—smells that compelled the princess to move, to follow, to run as fast as possible. She joined a stream of other princesses, flowing down corridors like living rivers, running on floors, walls, and ceilings to move as fast as possible.

And now something new. Light! She was running toward the light, and her eyes could now see her companions: dark shapes moving, running. As the light increased, the heat increased, and the muggy, moist taste of the air increased. She felt as though she was being sucked out of the tunnels, the drive to move forward was so great. At last she emerged into the blinding sunshine, blazing heat, and humid air. With her companions around her, she instantly spread her wings and flew, joining the cloud of princesses and princes that filled the air.

Like leaves caught in a miniature funnel of air, they spiraled up and away from the nest. Dancing with perfect choreography, they flew in a self-destructive orgy of relentless reproduction. The princess, compelled by the near sensory overload of new sights and odors, found her partner and mated in flight—then found another, and another. Each time, she stored their sperm within her.

At last, the clouds began to dissipate, and the princess flew away, leaving her exhausted mates to die. Her flight was now unplanned and largely random, but chance took her to an inviting cavern in the ground, where she crawled in, excavated a little, and rested.

The most important decision of her life made, she then became still. She did not notice or care when her once glorious wings fell from her back as the leaves fell from the trees around her. She simply rested as winter came and went.

As spring arrived, she awoke and lay her first eggs, fertilizing them with the sperm stored from her mating flight the previous summer. Her female larvae hatched after three or four weeks, and the fully fledged queen then spent the next three weeks feeding them from her salivary glands, making use of her extensive fat reserves built up during her own larval stage. Two weeks later, the queen greeted the first workers to join the nest as they emerged from their cocoons.

From that day onward, the queen was looked after by her daughters. Moving sandy soil grain by grain, those first workers excavated the beginnings of the complex nest that was to grow year after year, and led the queen to the safety of a lifetime in total darkness. The queen settled down to her life of egg laying and being fed. Despite her royal title, her role was now simply as an egg-producing factory. While she did her duty, her daughters—the workers—built, explored, nursed, cultivated, and protected.

Some seventeen years later, the nest was still there, now a vast underground city of throughways, food chambers, egg chambers, nurseries, and, of course, the queen's chamber. The design of the city was efficient yet mazelike. Like the organs of a larger creature, every part of the nest worked to keep itself alive. Workers constantly repaired and excavated new areas, creating mounds of sandy soil around the main entrances. Internal temperature was maintained by blocking or unblocking different openings. The nest was protected by workers that stayed by the entrances to defend against predators.

Every day, workers would venture out to gather food for the city. The nest now had an extensive territory, marked by pheromones. In certain parts of this area, the workers cultivated large numbers of aphids. These were milked for the honeydew they excreted, and in return the workers would protect these specific aphids from attack by predators. A certain kind of wasp, looking very much like one of the aphids, sometimes took advantage of this relationship, even enticing the workers to feed it.

Whatever they did, the workers always acted as a unit. If a big obstacle needed to be scaled, they would form bridges with their own bodies, allowing their companions to travel across. If a large piece of food needed to be moved, they would work together to lift it. If one found a new source of food, it would inform others, and the shortest route from the nest to that food would be used. If they had a choice of many different food sources, they would always use the richest, most nutritious food for the city. The workers even took care of their own dead, removing them and carefully placing them in a special area. It had become an impressive city, an extensive society.

The reign of our queen had been mighty and unchallenged. But this was to change. The queen's last hours were spent in her city on a hot and humid July afternoon. As always at this time of year, she had recently laid many unfertilized eggs in order to generate winged sons ready for mating. In the nurseries, workers had been rearing her daughters to be queens for the same event. Just as the swarms of excited princesses and princes flew out to mate, our seventeen-year-old queen died. Her workers, her daughters, on finding her dead, unceremoniously carried her to the area where all their dead were placed.

And thus ended the future of the city. Although the activity of the many thousands of workers continued unchanged beneath the swarming royalty in the air above, there could be no more workers. The nest continued its efficient, relentless tasks, but without new workers, all it could do was slowly wither like a plant with its roots cut. The queen may not have been the brains of the city—indeed, like most royalty, she was not in control of anything—but the city comprised her children, and without her they were doomed.

Overhead, a daughter, and now a mated queen, flew a random flight toward an inviting crevice. With her lay the potential of the next generation.

INSECT INTELLIGENCE

The story above has been happening since before the *Tyrannosaurus rex* ruled the land. Although trampled underfoot, these creatures have outlasted the stegosaurus, the saber-toothed tiger, and the woolly mammoth. Social insects have been building cities since before we were living in caves. They've been cooperating, communicating, and solving difficult problems since before our predecessors even knew how to climb trees. They've been looking after themselves, collecting food, nursing their young, and removing their dead since before there were even such things as mammals.

Today there are so many different species of social insect that if you were religious you'd accuse God of extreme favoritism. Social insects are very successful. Extremely successful. *Staggeringly* successful. Take a walk through the streets of any city or indeed any bit of countryside during the hot, muggy days of midsummer and you will understand the point I'm trying to make here. Social insects will literally be underfoot and in the air wherever you go.

The story that opened this chapter was about a specific type of social insect: the ant. In particular, it was based on the behavior of the common black ant in Britain, *Lasius niger.* (It's my way of apologizing to the occupants of the nest I removed from my bathroom window frame last month.)

By looking through the eyes or feeling through the antennae of the ant, I was trying to show what a familiar and yet alien world ants inhabit. We are all so used to the idea of jobs, cities, and societies that these things seem quite natural to us when we hear of insects doing much the same things. Indeed, most of us have looked down from a tall building at rush-hour traffic and tiny people and thought, "They look just like ants!" But unlike us, ants have very little choice. An individual ant—even a queen, as we saw—is a very simple creature, driven almost entirely by instinct. Ants are little robots, programmed to do a small range of things.

This brings me to the other reason for the story: to make a specific point about intelligence and control. Despite the queen being called a queen, this has nothing to do with power, control, or intelligence. It is true that the queen will typically be much larger and live much longer than the workers in a nest (for *Lasius niger*, the queen may be 15 mm and live for nearly twenty years, while workers are 3–5 mm and live up to eight years). But apart from the fact that all new ants come from the queen, she has no real control over the day-to-day activities in the nest—not that she has the brain to be able to direct, rule, or order anything, anyway. Like the workers, she is a simple preprogrammed automaton, capable of only a few tasks. The name *queen* persists only because of historical reasons. We now understand that queen ants are not rulers, and there is no hierarchy of rank in insect society. Indeed, there are no ants cleverer than any others that could give orders and form plans.

How can this be true? In the story, we saw that ants build complicated cities, regulate temperatures, maintain nurseries, farm aphids, cooperate to carry things, teach each other where food is, find the shortest path from the food back home, and so on. And other ants will invade other nests, use chemical warfare, steal larvae, and then enslave the ants when they grow. As another example, honeybees have some uncanny abilities to decide where the best flowers are located and thus maximize their pollen gathering. And termites can build structures that make our best cathedrals look amateurish. Surely all of these impressive accomplishments need somebody in charge to direct the activities? Some boss to order everyone about?

When we build something, there's always someone (usually the one with the money) who decides when and where it gets built, and someone to design it, for that matter. When we go to war, we have generals to direct our armies. When we go out foraging for food (in the local supermarket) someone works out what we need and what we don't (usually the wife or girlfriend).

But as we know, social insects are not really any cleverer than any other kind of insect. A single ant is stupid. Indeed, every single ant is stupid. How then can a colony of ants become clever?

To try to answer this particularly difficult question, we must stop thinking so hard about separate ants and think more about nests. You may have noticed a particular sentence in the story before: "Like the organs of a larger creature, every part of the nest worked to keep itself alive." In essence, this is how a collection of many stupid ants can create intelligence. Ants (and all other social insects) act as cells do in a higher-level organism. The individual cells are stupid, but the organism—or nest—is intelligent.

This should sound quite familiar after the previous chapter, where we saw that many simple cells with many interactions between them behave as a brain. But a colony of ants is not a brain. It is actually more than that; it is an entire organism.

NESTS AND ORGANISMS

As you read this, consider yourself. You are a vast collection of cells working in harmony to keep the collective alive. Your cells are in clumps, which we call organs. Each clump is typically made from specific cells specialized for different tasks. For example, your liver cells are grouped together and do "liver things" (such as secrete bile, store and filter blood, and convert sugar to glycogen). Your skin cells are grouped together in a big sheet that covers you and protects you from bacteria and viruses as well as keeping you cool (or warm), generating certain essential chemicals from sunlight, and so on. Your brain cells are grouped inside that big cavity behind your eyes, where they are currently—I hope—thinking about themselves as they read this.

We are a community. Each of us is the result of cooperation among millions of cells. Your body does not have that many different types of cells (around 250 in one recent estimate), but because there are rather a lot of them and because they are organized in a rather clever way, they somehow act as a single organism.

Now think about an ant nest. A nest comprises thousands of smaller, specialized things. Some maintain the temperature of the nest, others defend the nest from intruders, others go out and gather food, while inside these small things look after each other, remove the dead, and help the new ones to grow. Not cells, but ants.

Even the concept of reproduction seems strangely similar to that of a larger organism. Although a nest may comprise thousands of ants, they do not all reproduce. In fact, all reproduction takes place via the queen. She generates haploid males (developing from unfertilized eggs with half the usual amount of DNA), which die immediately after reproduction, just like the sperm of a larger organism. She also generates queens (this time with the full quota of DNA), which are very reminiscent of eggs from a larger organism. So you could think of the queen as being the reproductive organs of the nest organism.*

The nest also has a few other tricks up its sleeve—some nests permit more than one queen,† so even when the existing queen dies, this does not always mean the death of the colony as we heard in the story. A new queen (often a daughter of the original) can take over or even reign in parallel to her parent, allowing the life of the nest to be increased dramatically.

This "nest organism" still feels rather strange, though, for the concept of organs is only partially valid. Although there are different castes of ants just as there are different types of cells, ants are a bit cleverer than cells. So instead of over 200 different types of cell, most nests will not have more than four or five different types of ant. And instead of having the specialized cells clump together to form organs, a nest doesn't do much clumping. The eggs may be grouped in one area, the larvae in another, and so on, but generally, the organs of a nest are very *distributed* things. Moreover, because ants are more versatile than cells, they can be-

---

*I have to tread quite carefully here, because the dreaded words *group selection* are sometimes used in this context. Indeed, many opponents of group selection wrote off the "superorganism" argument because they felt that it relies on ideas of group selection to work. Thankfully, however, this no longer seems to be the accepted view. For those who know about these things, it now seems that the reproductive strategy of social insects can be explained by kin selection theories alone. Because of the increased degree of relatedness among workers in the nest, one can regard the nest as a multilevel extended phenotype of genes originating in the queen. Of course, there are complications of genetic conflict among different "selfish genes" here, but there is no reason that an evolutionarily stable state that consists of a subtle balance of the contradictory selection pressures for different genes should not exist in the nest organism. Similar balances (e.g., miotic drive) exist in organisms such as ourselves.

†It looks as though we can use kin selection to explain this too, according to Andrew Bourke, an entomologist at the Institute of Zoology in London.

have as more than one organ in the nest. So a collection of ants that were acting as "nest muscles" by moving food into the nest will also act as parts of the "nest brain" as decisions are made or part of the "nest metabolism" as temperature is maintained.

This is actually a rather clever thing to do if you're an organism. If you were being attacked, imagine how useful it would be to suddenly make all the cells in your stomach and intestine become muscles to make you stronger (the best our bodies can do is reduce blood flow to these organs and increase it to our muscles). Or imagine if you had a particularly tricky problem and could make all of the cells in your legs act as brain cells (assuming you were sitting down at the time). Clearly there are some very big advantages in being a nest organism. Indeed, this helps explain why these organisms have been around for so many millions of years.

So "nest organisms," or "superorganisms" if you want to use the original name, are rather clever things. But you may have noticed that I have not really answered that difficult question I posed earlier: How can a colony of ants become clever?

We know that a nest organism is clever while an ant is not, just as you are clever while one cell in your body is not. But how? Just what goes on when a group of stupid ants does something clever?

The answer comes from an unlikely source: physics and chemistry. Its name is self-organization.

SELF-ORGANIZATION

The ideas of self-organization were not created with social insects in mind, but they do help to explain a few things. The concepts were constructed to explain some strange observations in the world—unusual patterns in rocks, the highly regular structures of crystal growth, oscillating patterns in chemicals, and structure emerging suddenly at a certain temperature. These things looked as though a single entity was controlling them, designing them, but in reality they were *organizing themselves.*

How do they do it? There are quite a few different answers to this question. One common view is that a self-organizing system should be thought of as an *energy exchange system*. To use even more frightening terminology, self-organizing systems should also be *thermodynamically open*. What all this means is that there should be a flow of energy through the system—the system should be exchanging energy or mass with its environment. We can check that an ant nest does this: ants expend energy as they move themselves and objects in their environment; they gain energy from food sources in their environment.

The trouble with these energy exchange systems is that they like to either settle into a state of equilibrium or fly off into randomness. To stop a system from settling into a stable state, it needs to be dynamic—undergoing continuous change. But to stop it from becoming entirely random, that change should not become excessive. Because of this, it is often said that self-organization lies on the boundary between order and disorder, or "on the edge of chaos."

As an example, consider water. If there is insufficient energy and change in the molecules of the water, it settles into equilibrium—it solidifies into ice. But if there are too much energy and change, the molecules start to jiggle about randomly, and you've got water. For self-organization to occur—for example, patterns of frost on the window or the delicate structures of snowflakes—we need something halfway between the two.

You're probably thinking, "So we need energetic, snowy ants. But what has this to do with lots of stupid things becoming clever when they work together?" Well, you're quite right. Such concepts can be useful tools to help us think about self-organization (and we will return to ideas like these in later chapters). However, thinking about being on the edge of chaos does not really *explain* anything about our ants right now.

Thankfully, instead of having to rely on the exotic theories of mathematicians and computer scientists (who are usually the ones to blame for the previous ideas) we have some more down-to-earth ideas thought of by the researchers who actually work with social insects, and indeed by those who work with the digital equivalents inside our computers. I am talking about Nigel Franks (a professor of animal behavior and ecology, Bristol), Jean-Louis Deneubourg (a specialist in complex systems dynamics of social insects, Brussels), Thomas Seeley (professor of neurobiology and behavior of social insects, Cornell), and swarm intelligence pioneer Eric Bonabeau, among others. These guys really know their social insects, and not surprisingly, it is they who have the best explanations for self-organization in insect colonies.

To help me explain how it all works, let me give you a more concrete example of intelligent behavior by ants. Let's briefly return to the story of the nest that began this chapter. Around five years have gone by since our queen founded her city, and it is morning.

COLLECTIVE DECISION MAKING

The worker waited patiently as her sister unblocked the exit from the nest. She then scrambled through it, standing briefly on the sandy mound in the sunshine, antennae waving as she detected the different pheromones around her. As usual in the morning, the scents were fainter and less distinct. Nevertheless, she could still feel the pull of the "follow-me" scent laid down the day before.

Today she resisted the chemical instruction. Instead she took off in a new, completely random direction, running for a little while, then stopping to check the scents, running in a new direction, then stopping for another antenna wave. She proceeded in this meandering, staccato fashion for some time, her path a child's scribble on the ground. Often, during activities such as this, she might come across the main worker path, as full of traffic and scents as a highway is of vehicles and noise. On these occasions, she might simply join the traffic and go home, or help her sisters with their foraging.

This morning was different. Her wanderings had taken her to a relatively unexplored region, and she had found something new: aphids! Those delightful creatures provided such delicious honeydew when prompted by a prod and a squeeze in the right place. The worker explored the area some more and fed, quickly discovering that this new crop of aphids was extensive. She then headed home, using her memory and scents to guide her, leaving a pungent new trail of pheromones as she went.

Because there were now so many workers in the nest, she was not the only one to go exploring and not the only one to find this new, rich food source. Not long after she arrived back at the nest, her sisters were also returning from the same place, many having followed some or all of her trail home. As the pheromone on this new trail grew stronger, more ants were compelled to follow it instead of the existing routes. Soon the trail to the new source of food had become another busy highway, running a direct and efficient path from the nest to the food and back.

As usual, the nest had made its decision quickly. Faced with more than one source of food nearby, it picked the richer source, and made sure that its "muscles" used the most direct route to reach it, thus minimizing the effort it needed to expend in order to feed.

There we leave our ant nest again, for the time being. But I hope you now understand a little more of "nest intelligence." Faced with problems such as these, the nest makes collective decisions—choices and plans that are beyond the capabilities of a single ant.

"Oh, big deal," you're thinking. "So ants can find food." Well, actually the task just performed was rather clever. Here's a similar problem. Imagine you're on vacation in a strange city, in your hotel, and you decide to go out and eat at a restaurant. You have a number of choices; some will provide nicer food, and some will be a little cheaper. Which do you pick? Okay, now imagine you have to walk there and back, and you don't know the area. You need to find the best restaurant *and* you want to find the shortest route there, navigating through miles of complicated alternate routes. Unless you're prepared to eat at an inferior restaurant and get lost a few times on the way, it's going to take a bit of planning first. You're going to think for some time about different menus, and pore over a few maps.

Writing as someone who prides himself on having a good sense of direction and a reasonable intelligence, I can truthfully say that on most occasions, I am not terribly good at solving this problem. I usually get lost a number of times and end up finding the nearest place to eat and getting a cab back. I suspect I am not the only one, too. This is not a trivial problem.

And yet an ant nest can solve this problem. Not only that, nests solve the same problem every day, and many other similar problems too. Indeed, they solve this kind of problem so well that, as we shall see later on, we use digital equivalents in computers to solve massively complicated scaled-up versions of the same problem.

But for now: back to the point. The nest was intelligent. It made a choice, and it performed something that resembled planning. This happened because it did four important things. First, it used *multiple interactions*. Second, it used *positive feedback*. Third, it used *negative feedback*. Fourth, it *amplified fluctuations*. Let's work out what these four processes are in more detail before we see how the nest uses them.

Multiple Interactions

After the last chapter, the idea of multiple interactions should need very little explanation. Essentially, we need lots of individuals who interact with each other in some way. Actually, some form of self-organization and collective decision making can occur when a single individual interacts with itself or with the results of its own activities. Generally, however, we need lots of them.

An easy example of this is a conversation. When we talk to one or more other people, what each of us says next will depend on what the others have just said and on what we have just said as well. It is the multiple interactions of a conversation that makes it so dynamic and interesting. Without such interactions, we would all simply go around having monologues.

Positive Feedback

In a word, positive feedback means amplification. But to explain further, let me give you an example. No, not the irritating whine of a speaker placed too close to its microphone. Food.

When we eat snacks such as a packet of chips or a chocolate bar, we experience positive feedback. Every time we finish one, we want another. These foods are designed to make us want more. And this is what positive feedback is all about: the act of doing something causes the same thing to happen again. So the next time you find you've eaten a huge packet of chocolates, you can blame positive feedback.

Negative Feedback

If all we had was positive feedback, things could get a little messy. Thankfully, there is a counterbalance or brake: negative feedback. To explain, let's consider another example. No, not steam engines. Food. Again.

Although we could go on eating snacks forever if positive feedback got the better of us, we do tend to experience negative feedback. When we've eaten enough, our appetite decreases. If we still persist, our stomach begins to become uncomfortably full, and we may even begin to feel a little nauseous. So we stop eating. This is negative feedback: the act of doing something makes it less likely for the same thing to happen again. When our negative feedback is nicely balanced with our positive feedback, it ensures that we eat just enough, but not too much.

Amplified Fluctuations

The fourth and final process that self-organizing systems need is the ability to amplify fluctuations. You may have heard of the butterfly effect, where a butterfly flapping its wings on one continent causes a hurricane on another continent. Or the story of the kingdom being lost because of the lack of a nail for a horse's hoof. This is the same idea, but let me give you an illustration that is a little closer to home.

Britain has a national lottery. Balls are randomly picked by a machine twice a week, and if the numbers on your ticket match those on the little rubber balls, you win. Many thousands of people around the world have become millionaires through this kind of game. Their lives and the lives of people around them are changed in dramatic ways—sometimes good, sometimes bad.

This is an example of amplified fluctuations. In the normal course of your day, you really should not care if a few rubber balls are picked out by a machine. Let me do something similar now: I'm putting my hand into my desk drawer without looking, and I've picked up a hole punch. Not really very interesting, is it? But because the result of the rather uninteresting random "rubber ball event" is *amplified* to the extent that it could change your life, it becomes significant.

So that's what was going on in the ant nest earlier: multiple interactions, positive feedback, negative feedback, and the amplification of fluctuations. Remember when the worker decided not to follow the normal trail and instead went exploring? On average, around 10 percent of ants make this decision entirely randomly. But because the worker went exploring, she found the new food source. And because she and others found the new food source, the nest eventually "chose" to redirect most of its activities from an existing source to the new source. This was an amplification of a fluctuation. Without that fluctuation, the nest could only continue to do what it was currently doing. The randomness created new opportunities and allowed new choices.

So random fluctuations in behavior provide the nest with alternatives to amplify, or choose. But how does it choose?

If you recall, our worker laid down a strong pheromone trail from the food to the nest as she returned. This had the effect of encouraging other ants to follow her trail, find the food, and lay down their own pheromone trails. The more ants that followed the trail, the more trail there was to follow, so the more ants followed the trail. This positive feedback is how the nest amplified the good fluctuation and made its choice. And the positive feedback was caused by multiple interactions among the ants as they sensed existing pheromone and laid down their own.

Finally, negative feedback was inherent in this system. Because there were a limited number of ants and a limited amount of food, the number of ants following the new trail to the aphids could not continue to increase indefinitely. Once the number of ants reached a certain level, the amount of food would start to decrease too quickly, with demand outstripping supply. Then some ants would not find this source of food quite so tempting and would not put down more pheromones, reverting instead to the old trail and allowing a balance to be maintained between this and the existing food sources.

A similar process was followed to enable the nest to find the shortest path to the food, but we shall examine this in more detail later. For now, just be content with the knowledge that you know how a collection of stupid ants can collectively make intelligent decisions. They use multiple interactions, positive feedback, negative feedback, and the amplification of fluctuations.

Ants are not the only insects to do this. When termites build their extraordinary pillars and domes, all they do is follow two rules: move toward the strongest pheromone, and deposit what they are carrying where the smell is strongest. The multiple interactions, positive feedback, negative feedback, and amplification of fluctuations enable their impressive architecture to emerge. When bees choose among sources of nectar, their choice is made in the same way. And the organized patterns on the combs of honeybee colonies are generated using the same four processes. Indeed, these concepts are sufficiently general to be applicable to many types of self-organizing systems, from social insects to crystal growth to cloud formation. And in the digital universes of our computers, we have been using these ideas for many years.

AGENTS CAN BE USEFUL

There is a field of computer science that does nothing except try to make agents do useful things. (If you have a real estate agent or literary agent, you will know exactly how hard that can be sometimes.) The field is called *distributed artificial intelligence* (DAI). The DAI agents in our digital universes are not quite the same things as the agents in our universe. Inside our computers, an agent is an entity, distinct from its environment, that has some kind of behavior.

You'll notice how careful I was in that definition and also how general. The trouble is, the word *agent* is often used to mean a lot of things, and there are a great deal of definitions. Here's another: "An agent is something that collects information and performs actions." Or how about this: "Agents are machines or beings that act in a world." Or perhaps: "An agent is a computational process with a single locus of control and/or intention."

But although we find it hard to agree on exactly what an agent is, we have been using them for many years to solve a diverse range of problems. As usual I have no wish to try your patience by waffling on at length about every agent-based application. I also have limited space, so here's a few applications I won't be talking about: diagnosis, scheduling, learning, negotiation, electronic commerce, reasoning, planning, game playing, speech understanding, air traffic control, electricity distribution management, and many, many more. Like social insects, DAI is very widespread and can be found where you least expect it. For example, the chances are that your favourite Internet search engine sends out little agents to find your request every time you search for something.

These agents are, of course, created by us. They are little digital pets, defined and constructed out of separate little computer programs. Also, they are often a little cleverer than the ants in the ant colony we have seen so far. These agents typically reason a little bit, make and store plans, construct internal world models, or communicate using words. Although inspired by social behavior in nature, DAI often relies upon fewer, more intelligent agents, who hold little symbols in their digital brains. They might remember things like:

```
"<I> next to <AGENT2>"
```
and
```
"<I> carrying <FOOD>".
```

The functionality of all of the agents then becomes a result of symbol processing between them. For example, we can enable agents to pass objects between each other by applying the rule:

```
"if <AGENTx> gives <OBJECT> to <AGENTy> then <AGENTx>
is not carrying <OBJECT> and <AGENTy> is carrying
<OBJECT>"
```

Although this can be very effective, it is often *explicit functionality*. We have carefully designed all of the agents to make them work together and do the right thing. Rather than trying to use the feedback and amplification processes we saw earlier, these agents often rely on multiple interactions alone. Consequently, the most important and difficult problem for researchers in this field is trying to make all of the agents work together by designing complicated coordination and communication systems.

However, ideas are changing. As we learn more about how social insects such as ants achieve their impressive feats, the more traditional ideas of symbol processing and planning are being replaced with newer, more natural methods.

EMERGENT FUNCTIONALITY AND COLLECTIVE ROBOTICS

Symbol processing is less in vogue than in the past. With the increase in understanding of hive intelligence, many computer scientists use a different kind of agent-based approach. Larger numbers of less intelligent agents are used. Agents now think *subsymbolically*. If they remember anything at all, it is stored using their own digital neurons or some equivalent, not using lists of symbols. And those four important processes are now the main reason for the functionality of the digital agents, just as in social insects.

Perhaps the most visible demonstrations of this change in thinking are in the field of robotics. Instead of concentrating on one or two very intelligent robots, many research centers around the world are creating little "nests" of many less intelligent robots. The robots work together, following very simple behaviors but using multiple interactions, positive feedback, negative feedback, and the amplification of fluctuations to achieve their goals.

Let's look at a specific example: the puck-sorting robots of Ralph Beckers, Owen Holland, and Jean-Louis Deneubourg (also duplicated by Rod Goodman and his colleagues at Caltech). These robots were actually even more stupid than ants. They could move about, see objects with their infrared eyes (such as other robots, walls, or pucks), and detect when they were pushing something heavy (another robot, a wall, or more than three pucks). And that's all.

Their behavior was equally simple. Each robot followed three rules. First, if it could see a wall, it would turn away. Second, if it was pushing something heavy, then it would reverse and make a random turn. Third, if not pushing something heavy and it couldn't see a wall, it would go forward.

When placed in an arena surrounded by walls with randomly scattered pucks, the robots would trundle about, looking like ants as they followed their rules. Very quickly, an emergent behavior would become evident: all of the pucks would be clustered in one area. The very stupid robots, following their extremely simple rules, managed to push all of the pucks into a single group. And with only a minor modification to the second rule (pull the puck back a distance, depending on its color), the robots suddenly begin sorting differently colored pucks into different piles.

And it was done—you've guessed it—by using those four processes. There were many fluctuations as each robot wandered around in the arena, its random path dictated by the rules. By chance, each robot would start to push pucks in front of itself, and also by chance, sometimes a robot would manage to push three together, at which point it would turn and go elsewhere (rule two). With three pucks together, it was then more likely for the robots to push other pucks into the growing pile. As the pile grew, so the likelihood for it to grow increased—positive feedback, amplifying that initial fluctuation. The multiple interactions were caused by each robot affecting the environment of the other robots by pushing the pucks around. The negative feedback was simple: eventually they ran out of pucks, so the pile could grow no more. It's another example of emergent intelligence arising from many stupid things working together. And this is exactly the process ants use as they sort eggs from larvae and remove their dead to a specific location.

EVOLUTION AND NANOROBOTS

Not only are we now using emergent functionality of multiple robots, we are also using evolution to generate those simple behavioral rules that create the emergent functionality. A rather nice example was shown recently by Jordan Pollack and his team at Brandeis University. They allowed their collection of simple robots to *breed* in their environment. Although no "baby robots" were created, they did permit new robot "brains" to be generated by gene transference from parents into other robots. (The robots transmitted their digital genes to each other via infrared signals.) Energy levels were used to provide the selection pressure for the evolving robots. Those robots that managed to move closer to a light source got more energy. Those that couldn't work out how to move into the light gradually lost energy. The robots with the most energy reproduced more, creating (via genetic crossover and mutation) and transplanting new brains into the robots with the least energy.

In this example, the robots evolved to find the light source in their environment, not a task that required collective intelligence. Nevertheless, such "embodied evolutionary" techniques seem ideally suited for the automatic generation of emergent behaviors by collective robots. Indeed, these researchers claim in their research papers that such approaches "have the potential to scale to very large systems (on the order of hundreds or thousands of robots)" and predict that such robotic systems will become increasingly prevalent.

Other researchers have similar visions: "In the near future as the costs of robots are going down and the robots are getting more compact, more capable and more flexible, we expect to see industrial applications of Very Large Scale Robotic (VLSR) systems, for example, many thousands of mobile robots performing tasks such as assembling, transporting, and cleaning within the working space of factories." Some of the predicted applications are less pleasant: "VLSR systems consisting of small crawly robots can serve in the army as spies for inspecting perilous areas."

Which brings us to yet another, rather remarkable avenue of research. Nanorobots.

Research in this area is still in the what-if stage, but we'd like to be able to construct and use thousands of molecular-sized robots, capable of building and repairing our technology and even ourselves. Although still a dream, the idea of a swarm of collectively intelligent robots inside our veins working to keep us alive is something that some researchers are actively working toward. There is still a long way to go, with no clear answers to even the most fundamental questions such as "How do we build them?" but researchers have shown that we can make microscopic gears and use DNA to generate molecular-size structures with specific shapes. Some even suggest that DNA motors are possible, capable of propelling nanomachines along. And the proponents of this future technology are enthusiastic, quoting benefits such as enormous computation power and vast memories created by a swarm of so many thousands of nanorobots. Even self-assembling nanorobots are being discussed by the theorists.

No one really knows whether any of this will ever happen, but nature has got the closest so far. As we will see in the chapters on growth and immune systems, our bodies already contain cells, viruses, and proteins that build and repair us. These are nature's nanorobots, with their own collective intelligences.

But having ventured out of science and into science fiction, let us return to what we can do today. You may have noticed that although this chapter is entitled "Insects," I have been talking about ants quite a bit. There is a reason for this. There just happens to be a new field in computer science called ant colony optimization.

ANTS AND PATHS

Earlier in this chapter, we saw a worker and her sisters discover a new food source, and we worked out how the nest chose to exploit this food rather than the existing food source. At the time I mentioned that as well as making this choice, a similar process was followed to enable the nest to find the shortest path to the food.

It's true. Nests are very good at making choices, not just about which food source to use but also about which path to take in order to get there. This is important if you're a nest. Your hundreds of workers will all follow the same path to the food and back, so if that path is even a tiny bit longer and less direct than it needs to be, you're causing those hundreds of workers to waste energy. It's in the nest's best interests to find the shortest path from itself to every new food source. And so that's exactly what it does.

We already know the mechanism used by the ants: the buildup of pheromone trails using those four processes. But to explain, imagine this:

On her way to the new food source, a worker discovers that she has to cross a patch of water. There are two branches bridging the water, so she can take one of two paths. She picks the left one randomly. Behind her, another worker is faced with the same choice. This worker randomly picks the right-hand branch.

It turns out that the left-hand branch was a rather long and twisty one, and so made the journey across the water twice as long as the branch on the right. This meant that in the time it took one ant to travel down the left-hand branch, another ant was able to travel down the right-hand branch to the food and back again. Because the ants were laying down pheromone trails as they scurried along, this meant that the right-hand branch got more pheromone deposited on it than the left-hand one. With more pheromone on it, more ants were compelled to use the right-hand branch, and so more pheromone was laid down, and so more used this branch, and so on. The familiar process of positive feedback allowed the nest to choose the shortest branch.

On every trip to a new food source, there are countless choices to be made like this, and the nest picks the shortest route each time using exactly this approach.

In computer science, there is a type of problem known as the *traveling salesman* problem. Imagine you are a salesman who has to travel between some cities. You need to find a route that will minimize the travel time (and cost) while allowing you to visit every city. Sounds easy? Well, for three or four cities, it is. But imagine you have to visit ten. Or fifteen. Or twenty. The number of possible routes you can take rises exponentially, making this an extremely hard problem to solve for a large number of cities.

A few years ago, a Ph.D. student named Marco Dorigo connected the dots. He saw how clever ant colonies are at finding shortest paths and saw how useful it would be to solve traveling salesmen problems. His doctoral work, in which he used digital "ants" in his computer to solve these problems, began the new research field called ant colony optimization.

Marco's approach was rather clever. He created a digital universe consisting of a set of cities and all the possible paths between them. He then deposited agents (his digital ants) that wandered around this network, laying down their digital pheromone trails. After a number of "ants" had managed to complete an entire tour by visiting all of the cities, the "ant" that found the shortest tour would strengthen its pheromone along that route (equivalent to the effect a larger number of ants would have). The digital ants would then continue to wander around the network but would become attracted to the shorter path more and more, strengthening it by laying down more pheromone, or finding even shorter variations, which then became even more popular. Using the same ideas of multiple interactions, positive feedback, negative feedback, and the amplification of fluctuations, the digital ants found the shortest path between cities, just as the ants in nature find the shortest paths from nest to food.

Although this field of computer science is still quite new, the number of applications is already impressive. Ant colony optimization and similar approaches have now been used for adaptive routing on communication networks (i.e., finding the best way for one computer on the Internet to talk to another), automatic vehicle control, scheduling, learning, function optimization, electronic circuit design, and many other applications. Researchers have also combined ideas from ant colony optimization with evolutionary search, using a combination of evolution and the distributed decision making of ants to solve engineering design problems.

And a new research project called DREAM (Distributed Resource Evolutionary Algorithm Machine) has even more ambitious goals. The researchers on this large, collaborative European universities project plan to allow the creation of DREAMs using computers connected via the Internet. Each DREAM will "support a number of infoworlds, each with an evolving population of infohabitants." The researchers plan to test this extensive digital universe by applying the digital ecologies within it to applications such as scheduling, data mining, and the simulation of complex economic and social behaviors.

But enough of ants. There is one final type of insect intelligence that we should explore. This one will probably surprise you.

SWARMING BOIDS

At the start of this chapter, we saw a swarm of flying ants reproducing in the midsummer sunshine. We're all quite used to seeing swarms of insects behaving in this way: a living, dancing cloud that turns itself inside out and folds in on itself as it moves through the air, avoiding obstacles such as ourselves. When locusts invade, the effect is even more remarkable. The sky becomes dark, filled with purposeful smoke, an enormous organism able to reach out and smother bushes, trees, entire fields with its tendrils. The coordination of the separate insects within such swarms seems to defy possibility.

This phenomenon is not unique to insects. Schools of fish, herds of buffalo, flocks of birds, even crowds of commuters can behave in the same way. They somehow seem to move as one, flowing like rivers, yet suddenly changing direction as though choreographed. There's something very simple yet very clever about all of them. It's called swarm intelligence.

SWARM INTELLIGENCE

It seems that social creatures, from ants to antelopes, have something in common: they like to be in groups.

This may seem particularly obvious, but it is important. To social creatures, a group is attractive. Indeed, the attractiveness of a group to an individual has even been measured. In a school of fish, it seems that larger schools are more attractive than smaller schools, and also the rate of increase of attractiveness decreases as the school gets bigger.

To understand what this means, imagine you're in a hurry, walking down the street, and you see a couple of friends on the other side of the road. You'll probably just wave and shout, "Hi! Catch you later!" Now imagine that you see a group of ten friends. You're much more likely to stop and find out what's going on. Bigger groups are more attractive. Now imagine the difference between seeing two friends and three. The relative change is quite large; indeed, it may be enough to make you stop and talk. But imagine the difference between seeing ten friends and eleven. There's not much change there. So although larger groups are more attractive than smaller ones, as the size increases, the rate of increase of attractiveness decreases.*

So far, swarm intelligence seems rather easy. After all, we're social animals. A crowd in a playground quickly grows—groups are attractive. And the other element of swarm intelligence is even easier.

When you're in a city during rush hour, it's difficult to move without other people getting in your way. You have to adjust your speed and direction, constantly trying to avoid collision. And most of the time, we don't violently bump into other people. Even in crowds of tens of thousands, all moving along together, the number of severe collisions is surprisingly low—which is the way we like it. We don't enjoy colliding with other people, just as birds don't enjoy flying into other birds and fish don't enjoy swimming into other fish. We all like to maintain a certain distance between ourselves and the others around us. On the highway, this can be so important to us that when the right distance is not maintained, road rage can result. Normally, however, if someone in the car ahead starts to slow down for no apparent reason, it will simply cause a ripple effect, with many other cars behind slowing down or adjusting their direction.†

---

*Does anything about this growth of groups sound familiar? It should: it's another example of positive and negative feedback.

†These ideas should also sound familiar. They are more examples of multiple interactions and amplification of fluctuations.

So we've got attractive groups, and noncolliding things in the groups. And believe it or not, that's all we need for swarm intelligence. The groups are formed and maintained by the attractiveness rule; they move in such unity and with such coordination because of the noncolliding rule.

Examples of swarm intelligence are easy to find. You've no doubt seen schools of fish on natural history programs swimming and turning from a predator as one, as synchronized as the slats in a venetian blind. This is swarm intelligence in action, enabling a group to make sensible decisions almost instantaneously. How do they do it? In fact, the fish do not all turn at the same time. Their reactions are so fast that it looks as though they are happening together, but in reality the change of direction ripples through the school.

What happens is this. The fish at the front see the predator and turn away. In doing so, they get too close to the fish behind them, which turn away to avoid them, and so get too close to the fish behind them, and so on. Like the chain of cars that must slow down on a highway if someone ahead brakes too fast, the fish all influence each other. So although the fish at the back of the school cannot see the predator they are fleeing, they will turn away simply to avoid swimming into their companions. And the fish do not scatter in all directions because of the attractiveness of the group. If any fish strays too far from the edges, it will instinctively swim toward the center of the group again.

So the two rules of group attractiveness and noncollision cause those now-familiar ideas of distributed decision making and self-organization.

BOIDS AND BRAINS

Craig Reynolds, a specialist in 3-D computer animation, performed what was perhaps the most influential work in computer science on this subject. He created a digital universe in which agents, or *boids* as he called them, flocked according to three rules. First, try to match the speed of your companions. Second, try to move toward the center of the flock. Third, avoid colliding with any of your companions. The results were stunning: a flock of boids that behaved in exactly the same way as a flock of birds or a swarm of bees. They swirled together, twisted and turned together. They had swarm intelligence.

Craig went on to help create animations of flocks and herds in movies such as *The Lion King*. (Plate 7 shows an example of digital fish by Bill Kraus that swim in schools according to Craig's algorithm.) But in computer science, other uses were being thought of for swarm intelligence.

To explain how we have harnessed this new form of intelligence in our computers, I must first tell you of a new type of digital universe.

In computer science, we often think about solving problems in an interesting way. We create a digital universe for a problem, which contains every possible solution to that problem. A solution in such a place is defined by its location in the space. So if I'm standing at point A in this space, then I can see one solution. If I move to point B, I'll see another solution, and so on.

It's not at all hard to create such spaces. Remember the Venus flytrap problem in the Evolution chapter? We wanted to know how much water to give the plant each day in order to keep it alive. The problem had one parameter—the amount of water—and the answer would be a single number. By identifying the parameter and working out the range of possible values for that parameter, we define the space of all possible solutions. So we might say that our parameter is *amount of water* and the possible values range from zero to *top of watering can*. Our universe would then contain possible solutions such as: empty, one drip, a third full, half full, two-thirds full, three-quarters full, and so on.

In this case, with a single parameter, the space has a single dimension. Each solution is a point on a line. If we had three parameters, then it would be a three-dimensional space. If we had five hundred, it would be a five-hundred-dimensional space.

These strange digital universes are called *search spaces*. Having created the space, we use the computer to search for a good solution. This is often how we enable computers to solve problems: we define them as search problems and use various methods to find the solution.

We already heard about one type of search algorithm in Chapter 3: the evolutionary algorithm. Evolution can be said to search for good solutions in the search space (although as we've seen, this idea is not necessary to understand evolution). But what about swarms?

Well, this is the clever bit. Researchers such as Russ Eberhart and his colleagues at Purdue University thought of creating a search space and then allowing agents or boids to swarm within that space. As the agents flew around, they would be searching the space for new solutions (remember that every location in the space corresponds to a solution to the problem).

The agents were very simple (indeed, so simple that Russ dislikes even using the word *agent* to describe them). They were attracted to the center of the swarm, and—like birds attracted to their roost—they were attracted to areas in the search space that contained better solutions.

Their search spaces were created for problems to do with neural networks. For example, in one of their earliest experiments, Russ decided to use a feedforward neural network to act as an exclusive OR logic gate.* Instead of using the traditional backpropagation algorithm to fine-tune the weights of the neural net (see the previous chapter), they used a swarm. They created a search space that contained every possible combination of weights for the neural net and added some randomly traveling agents or boids. Because the speed of every agent was influenced by both its companions and the need to travel toward better solutions, the swarm acted as an intelligent unit. It explored the space, pulling disparate parts of itself toward better solutions as they were found.

It turned out that swarms really are intelligent. Swarms are very good at finding solutions to hard problems and finding them *fast*. In a later experiment to find the weights for a neural network used to control a battery recharger, the swarm found the solution in 2.5 minutes, compared to the traditional backpropagation algorithm, which took 3.5 hours.

Although still very new, these techniques (sometimes also called particle swarm optimization) are showing impressive results in many application areas. Much work on fine-tuning neural networks has been performed, as well as ingredient mix optimization, reactive power and voltage control for a Japanese electric company, and visualization of hard-to-read data.

Chris Langton, often regarded as the founder of the field of artificial life and president of the Swarm Development Group, has formed Swarm

---

*An exclusive OR gate outputs a "high" signal when at least one of its inputs is "high," and a "low" signal when all inputs are "high" or all are "low."

Corporation, which applies these ideas to the problems of other businesses. The organization has released its software (called—wait for it—Swarm) under a GNU license, which means that it is available free, rather like the Linux operating system for computers.

In a similar vein, researchers at British Telecom's Future Technology Laboratory (now called BT Exact) have created software called Eos, which allows digital ecosystems and evolution to be used for applications such as aerial placement for mobile telephones.

Like ant colony optimization, the use of swarms to solve problems looks like another winner.

SUMMARY

Insects get a pretty tough deal. Nasty creepy crawlies that get into your homes or fly into your hair. Stupid, irritating, unpleasant.

That's what you may have thought of them. It's the reason we began by spending some time looking through the eyes of an ant as she grew up and founded her own nest. By seeing her life and the life of her daughters from their perspective, I hope I have opened your eyes to the other side of insects.

Like all of the natural world examined in this book, insects are beautiful. They are nature's robots—individually unintelligent, collectively brilliant.

This chapter has explored this other form of intelligence. An intelligence of a higher-level organism that arises from a collection of smaller insects. Collective decision making by a nest on the edge of chaos as it chooses between sources of food. And all of the self-organization arising from four simple processes: multiple interactions, positive feedback, negative feedback, and the amplification of fluctuations.

Our digital universes have become colonized by these creatures. DAI agents, collective robots, and even nanorobots demonstrate how we are using the same processes seen in ant colonies with our technology.

And new developments in this area are still emerging all the time. We now use the methods of shortest path finding in ants to enable digital ants to find the best routes for traveling salesmen or communication networks. We also use the strange intelligence of swarms in computers, using digital swarms to help design digital brains.

So the next time you see an ant nest, or a swarm of flies or flock of birds, remember that they are just as clever as you. They just think in their own distributed way.

Which is more than plants do. But although plants may not be intelligent, they do have structure. And as we shall see in the next chapter, some structure is so clever it might as well be intelligent.

6

PLANTS

Living glass snowflakes. Smaller than a pinhead, yet visible from space. Indiscernible to the naked eye, but sufficient to feed whales. Diatoms.

They are microscopic plants that make up a large proportion of the plankton in our seas. Maybe they don't sound much like the plants we're used to seeing every day, but then we don't see diatoms every day. In fact, we don't see diatoms at all without a microscope. But they are still plants. Although each one comprises only a single cell, they, like the plants we are so familiar with, use photosynthesis to help create their energy and grow. And when the plankton blooms, huge stretches of ocean become stained olive-brown by billions of these plants. Their sheer numbers mean that they provide sufficient food to support the largest creatures ever to exist on this planet, whales.

But size or number is not why I mention diatoms. We're starting this chapter with the humble diatom because of the way they look.

I'm looking at a picture of a diatom right now. It's a eucentric diatom, which means that it's round. In fact, it looks rather like a microscopic flying saucer. This one has spines that radiate from its center to the edges. There's another with indentations that make it resemble a tiny hubcap. And here's one that looks exactly like a fat coin, with thousands of tiny raised bumps around the edge and a strange symmetrical "head" in the middle (see plate 8 for some examples).

Each type of diatom has its own shell, or frustule, with a specific shape and pattern. Some are like the ones I have described; others are cuboid, or ellipsoid, or have other strange shapes. Every one is covered in patterns—

spines, bumps, lumps, dents—all beautifully arranged in rows and circles to produce some of the most stunning forms in nature. The forms are so complex that new microscopes are commonly demonstrated by taking ever-clearer pictures of them. The patterns are so attractive that compositions of diatoms have been made into collages by artists (see plate 8, left).

The snowflake-like forms are shells made from silica, similar to the stuff your windows are made from (that's glass, in case you were wondering). They are massively elaborated cellular walls, their structures allowing them to withstand environmental pressures and control their buoyancy in water. But the most important things here are the patterns. Why are there patterns at all? Why not just have random lumps and bumps? Where did all of these patterns come from?

For diatoms, the answer is a little unexpected. The patterns of their wonderful shells are caused by bubbles.

We all played with bubbles as children, and most of us have delighted in the way they wobble and float with their rainbow surfaces swirling. You may have also noticed something else: when two bubbles touch each other without popping, they form a straight, flat boundary between each other. Its flatness is caused by the pull of surface tension—the cohesive force of the water molecules clinging to each other as tightly as they can. (At this point I must pause in order to blow some bubbles. All in the aid of scientific accuracy, you understand.)*

Well, it seems that my bubble blowing is not terribly impressive, but this is what I've just seen. As you've heard, when two bubbles touch, a new (more or less) flat surface between them is formed. When I made three bubbles touch, I saw three flat surfaces, forming a three-dimensional Y shape. When four touched, I saw an H shape. And when (after many tries) I managed to get six bubbles of the right size all touching, a little bubble cube was formed in the middle. I even managed eight bubbles, which produced a number of different internal square bubbles. (Plate 9 shows the kinds of things bubbles can do.)

---

*If you wish to try this, I suggest using two wire loops and some liquid soap mixed with a little water. Blow a bubble with one loop, and then catch it with the other. You can then use a soapy loop to blow into the bubble and create companions that touch each other.

What my simple experiment demonstrates is pattern making using bubbles. Separate bubbles may be spherical, but together they form flat edges, cubes, Y shapes, and any other three-dimensional shape you'd like. Create a froth of bubbles, and the boundaries between them form regular, repeating patterns.* And if you blow bubbles in something that solidifies, then the straight boundaries between the bubbles are preserved, even when the bubbles themselves are no more.

This is what diatoms do. Their cell walls extrude liquid silica in very carefully controlled froths. Specific portions of these solidify and are built up into the extraordinary glass snowflakes that surround them. They may only be single-celled plants, but their bubble blowing is unmatched on the planet.

NATURAL PATTERNS

The patterns of the diatom frustule are not unique to nature. Look at any plant, and you will see some obvious patterns: the symmetry of petals, the repetition of branching of stalks, the spiraling of leaves up a stem, the interweaving of seeds in a pine cone. Plants and patterns go together like songs and music.

This chapter is about pattern formation in nature, both at the scale of the diatom and at the scale of the rain forest. Patterns, fractals, chaos—they're the building blocks of nature, chosen by evolution but not created by evolution. These things are properties of the physical laws that make up our universe.

All of life displays these kinds of patterns, not just plants. But in the same way that I used ants to illustrate nest intelligence, I shall use plants in this chapter to illustrate pattern formation.

RIDING PHYSICS

Before we explore some of these patterns and work out how and why they happen, we should think about what is generating the patterns. We know that evolution is the master of design, capable of fine-tuning the

---

*Don't believe me? Put a straw in your soapy mixture and blow. Now look at the bubble honeycomb you've just made.

shapes of life into every conceivable form. We know that evolution likes patterns—it uses them to create segmentation in caterpillars and repetition of ribs in skeletons. But we also know that we don't need evolution to have patterns. Physics will generate patterns in nonliving and nonevolving systems.

The patterns of touching bubbles were an example of this. The structure is created by the laws of physics (specifically, laws concerning attractive forces between molecules). Crystal growth is another example: crystals grow in geometric, cuboid shapes.* Some crystals grow in tree structures, looking spookily like the branches of living trees—the patterns of frost on our windows can provide us with nice examples of this. In the desert, sand dunes form regular, repeating shapes. Some cloud formations paint the sky with regular stripes. The giant red spot on Jupiter has been a constant, regular feature for centuries, despite the fact that it is really made from gases within a 270 mph storm. Patterns of rock erosion can create such geometric, artificial-looking forms that aliens have been blamed for their creation. When water flows, it meanders in a snakelike pattern. You can see this as water trickles down the slope of your bathtub and also in the path of every river on the planet. The laws of physics create patterns quite independently of life and evolution. They provide "order for free," as complex systems expert Stuart Kauffman likes to say.

And this causes quite a debate. Because physics likes to create patterns and because much of life is constructed from patterns, scientists such as Kauffman, mathematician Ian Stewart, and, famously, nineteenth-century professor of zoology D'Arcy Thompson believe that physics may play a more significant role than evolution in the creation of life. The argument is about what does the more work: Is evolution our true creator, or is it the laws of physics? The debate causes great excitement in the media as evolutionary biologists such as John Maynard Smith speak out for evolution and Stuart Kauffman advocates physics. As geneticist Steve Jones said, "The arguments go back to the beginning of population genetics, back to Darwin: How much is deter-

---

*Just take a look at salt or sugar grains with a magnifying glass, and see how regular their shapes are.

ministic and how much is stochastic? Everybody now believes that there is a balance between the two. But where the balance lies is extremely hard to point out."

So it seems that as with most other such debates, the true answer lies somewhere in the middle. Evolution could not shape life without physics. But physics on its own cannot shape life, it can only provide possibilities from which evolution chooses.

Just as a game of chess can't happen without the checkerboard, so evolution can't create life without the patterns of physics. But I like to think of the physics–evolution dichotomy as being rather like a horse and rider. Evolution rides its trusty steed of physics, providing constant guidance and directing it where to go. Physics will find its own way much of the time (and when evolution loosens its reins, it is free to show some of its own beauty, as we shall see with certain shells). If evolution wants to go in a certain direction but physics cannot, then the steed will balk, and evolution must either give up or find a different route. But unlike physics, evolution "knows" in which direction it needs to go. Evolution is the master; it directs, cajoles, and persuades the patterns of physics to perform as it desires.

We know this is true, even in life-forms such as the diatom, which seem to rely on the nonliving pattern formation of bubbles so much. For while the bubbles provide the mechanism used by the plants to generate their complicated frustules, the exact shape is determined by the genes of each diatom. Genetically similar diatoms have similar frustules. When the genes differ, the patterns of the frustules differ. Like all the patterns in life, physics provides the means, and evolution finds the way. The partnership of horse and rider: physics and evolution is a good one. It has resulted in you and me.

We've already looked at evolution in a previous chapter. Let's look a little closer at the kinds of patterns that the laws of physics provide and how they are used in nature. If the patterns of diatoms are caused by genes guiding bubble formation, how are other patterns caused? Why are leaves and stems placed in such orderly arrangements? Wouldn't random growth work just as well?

The answer seems to be a resounding no. To explain why, let's think about pine cones.

PINE CONES AND GOLDEN SECTIONS

Look at a pine cone, and you'll see a tear drop–shaped object with many small, hard leaflets protruding. Look closely at those leaflets, and you'll see that they are not arranged in rows or columns; they are arranged in *spirals*. Starting from the bottom, the leaflets spiral up the cone, creating that familiar organic look that we see so often in nature.

If you look at the pine cone end-on, from the bottom, you will notice that each leaflet is a slightly different size, smaller at the bottom and bigger at the top (see plate 10, left). The difference in size is caused by their growth: every time a new leaflet is formed, it starts as a tiny bud that grows in the middle, at the bottom. As they grow, they are pushed outward in a spiral pattern. It is possible to measure the angles between successive leaflets around the center of the cone. That angle is 137.5 degrees.

So what? Well, it turns out that this angle is rather special—so special that it is called the *golden angle* in mathematics.

This is where mathematicians get excited, for there are lots of special numbers that start to appear. We know that when we look at the pine cone end-on, the angle between successive leaflets is 137.5 degrees. Suppose we then measure the distance around the circumference of the cone from one leaflet to the next, and then from the second all the way round to the first again. We end up with two numbers that depend on the size of the cone. These two numbers also have a very special relationship to each other—they form the *golden ratio* or *golden section*.

I can now put on my magician's hat and perform a trick for you. Let's say that when you measured the shortest distance between the two leaflets around the cone, the number was 21 mm. Well—no sleight of hand—I can tell you that the larger distance (from the second leaflet all the way around the cone to the first) is almost exactly 34 mm. Perhaps you have a larger pine cone? If the smaller distance was 89 mm, then I can tell you that the larger distance is very close to 144 mm.

Okay, perhaps my magic skills won't win many entertainment awards. But let me tell you how it's done. The two distances between leaflets separated by the golden angle on a pine cone will always form the golden ratio, no matter how large or small the pine cone. Whatever the exact values of the numbers, if they form the golden ratio, then when you divide the first by the second, you will always get the same number—approximately 1.618. This number is a fundamental constant of nature, rather like pi (π). It is represented by the Greek letter phi (ϕ) and is often called the *golden mean* or *golden number*.

The ancient Greeks knew all about these numbers. Indeed, they felt that the golden section was so important that they designed their temples, vases, and art around these proportions. They felt that this ratio was a universal, harmonious constant that was pleasing to the eye. Many people still believe that in a perfectly proportioned human body, the ratio of navel height to total height will equal the golden section.

Maybe the Greeks were right, for the golden ratio does seem to appear everywhere in nature. Italian mathematician Leonardo of Pisa (who was nicknamed Fibonacci) spotted this in 1202 when he thought about rabbits. This is what he did: imagine you have a pair of newborn rabbits, one male and one female. They breed every month, and every time they breed, a new pair of rabbits is born (one male, one female). How many pairs of rabbits will you have after a year? To work it out, let's look at the total each month. At the end of the first month they mate, but there is still only one pair. At the end of the second month, a new pair is born, so we have two pairs. At the end of the third, the original produces another pair, so we have three. At the end of the fourth month, the original produces another pair, and the pair born two months ago produces another pair, so we have five:*

0 RR
1 RR
2 RR RR
3 RR RR RR
4 RR RR RR RR RR
5 RR RR RR RR RR RR RR RR

A pattern begins to emerge as we continue: 1, 1, 2, 3, 5, 8, 13, 21, 34, 55, 89, 144, 233, 377 . . . The sequence is known as the *Fibonacci series*. And these special numbers are also found in other unexpected places. Count the number of petals, or pairs of petals, in flowers, and it is very likely that you will find one of those numbers. The iris has three petals,

---

*If you keep on going, you will see that after twelve months we have 233 pairs of rabbits.

buttercups have five, some delphiniums have eight, corn marigolds have thirteen, asters have twenty-one, and daisies can have thirty-four, fifty-five, or eighty-nine petals.

If you're good at cracking codes, you should have noticed that each number in the Fibonacci series is made by adding the two previous numbers together. And if you divide any of these numbers with the next one in the sequence, you have an approximation of phi—the larger the numbers, the closer to the true value you get. We've found our golden ratios and golden number again.

So once we know the golden angle, it's like finding a treasure trove, for we quickly discover the golden ratio (or golden section) and the golden number (or golden mean). And with these numbers we can construct *golden ellipses* and *golden rectangles* that are proportioned according to the golden ratio. When we subdivide a golden rectangle into a square and rectangle with the same proportions, and subdivide that rectangle into a square and rectangle, and subdivide that one . . . and then draw a line around them . . . we discover a logarithmic spiral—the same spiral seen in thousands of different kinds of shells, seeds, and plant stems. Nature loves these golden numbers!

Mathematicians are usually in ecstasy at this point, but this book is not for mathematicians, it's for you. And right now you're probably wondering *why*? Why does nature like these golden numbers? What is so special about them?

To explain, I'll set you a little imaginary problem. You are a chef in a Spicy Meatballs restaurant. You need to fill a plate with meatballs without piling any on top of each other. But your restaurant manager (who doesn't like you very much) demands that you add your meatballs only to the center of the plate, one at a time.

So you have a go. You add one and then another, so that the first is pushed to the left. You add another in the same way, which pushes the first two to the left. You do the same again, and push the first three to the left. Quickly you find you have a line of meatballs on the plate, but you have not filled any of the rest of the surface. Your customer doesn't get many meatballs and complains.

A new plate arrives. This time you add one, add another so that the first is pushed to the left, then add one so that the second is pushed at 90 degrees to the first (upward), then add one so that the third is pushed to the left, and then add one so that the fourth is pushed upward. You quickly end up with a backward L shape of meatballs on the plate, but still lots of empty space with no meatballs. The customer is not happy.

After many tries, you eventually work out what you need to do. A new plate arrives. You add one meatball, then add another so that the first is pushed to the left. You then add another so that the second is pushed at 137.5 degrees to the first. You then add another so that the third is pushed at 137.5 degrees to the second. You then add another so that the fourth is pushed at 137.5 degrees to the third. And every time you add more meatballs in the middle, you make sure all of the others continue to be pushed out at the same angles. After some time, you have a plate completely full of meatballs, with no spaces. The customer is very pleased and also compliments you on the spiral patterns you have made (see plate 10, right).

In nature, there are as many different ways of growing leaves, petals, and seeds as there are ways of putting meatballs on a plate. Evolution could have chosen any of a million different patterns for the leaflets of the pine cone. But it chose to use the spiral patterns at 137.5 degrees from each other because this is the most efficient way of packing as many leaflets as possible on a cone—just as it is the most efficient way of placing as many meatballs as possible on a plate.* And it is the laws of physics that make this true. In the same way that the circumference of a circle is 2π times its radius, spirals of growing objects separated by the golden angle pack together most efficiently.

That's why nature loves these golden numbers. They represent the most efficient ways to pack seeds, produce stems, sprout leaves, and grow petals. After billions of years of fine-tuning by evolution, life tends to be very efficient.

FINGERPRINTS AND SNAIL SHELLS

Evolution is very careful to choose the patterns of physics that increase the survivability of organisms. But not all patterns have such dramatic

---

*Okay, so this is a simplification—it is the most efficient way only if you have spicy *growing* meatballs that start out small and grow bigger as they cool down.

effects, and so sometimes evolution loosens its reins and lets physics have a little fun. You have examples of this effect on your fingers and toes. Our fingerprints are complicated patterns that arise because of the horribly complicated chemical control process that determines how we grow from our genes (so complicated that I devote an entire chapter to growth). Evolution has determined that we need such patterns in order to enable us to grip objects without slipping. But evolution doesn't care too much what the precise patterns should be, so physics is free to be creative, resulting in unique fingerprint patterns for every one of us. Even identical twins, who have the same DNA, have slightly different fingerprints. Unfortunately, we very rarely marvel at the beauty of these patterns—their use for police identification seems to have somewhat stained their reputation.

The patterns on the shells of many molluscs are unarguably attractive, and once again, evolution does not seem to care too much about some of them. Many spend their lives buried in mud, so the pattern is never seen, rendering it irrelevant for survivability. This means that physics is free to try out its unlimited repertoire of patterns on these hidden shells, resulting in astonishing variation among members of the same species. And these patterns are not formed with bubbles or golden numbers this time, but with chemicals.

As we will see later, much of the growth process relies on chemical interactions between growing cells. And physics plays a highly significant role here, because certain chemicals like to form patterns.

It was Alan Turing, the mathematician who had a profound impact on computer science, as we saw in the Universes chapter, who first suggested how chemicals might be used for pattern formation during growth. In the early 1950s, Turing proposed that the chemical signals, or morphogens, released by developing cells might act in competition with each other. Chemicals might enhance or suppress each other, rather like the signals between neurons. He suggested that you need a chemical to undergo an autocatalytic reaction, which results in more of the same chemical being generated and also results in the generation of another chemical that inhibits the generation of the first chemical (another example of positive and negative feedback). If these two chemicals diffuse at different rates, Turing calculated, then stationary patterns of chemicals would emerge, like the paths of ants or weights of neural connections we have seen in previous chapters.

He was right. Today we use computer techniques known as cellular automata* to demonstrate the same processes within digital universes, and we see stripes, spots, circles, fingerprint-like designs, and every other conceivable pattern. Turing's work was seminal. Despite his being a mathematician, not a biologist, his ideas now provide the foundations of our understanding of many patterns in life. We now believe that the spots on giraffes, the stripes on zebras, the patterns of snail shells, the shapes on butterflies wings, the spacing of veins on leaves, and the color designs on petals may be created by the use of these reaction-diffusion chemical patterns.

But we are straying away from plants in our examination of natural patterns, so for the final type of pattern I'd like to show you, let's return to a familiar type of plant: the fern.

FERNS

Remember my fossil trilobite in the Evolution chapter? It died around 390 million years ago. If we could travel back in time and see what the Earth looked like then, we would see many unfamiliar forms of life. But we would recognize most of the few plants that existed then. They were ferns, and from the fossil record, we know that the shape of ferns has not changed much in the thousands of millennia since my trilobite lived.

Ferns are so ancient that they predate flowers and seeds. They reproduce using spores, which develop into tiny plantlets with only half the genetic material required. With the right kind of moist conditions, these plantlets (or gametophytes) release sperm cells, which swim from the male parts of one plantlet to the eggs held by the female part of another. When an egg becomes fertilized, it takes over the gametopyhte and develops into a new fern.

Sounds rather involved, doesn't it? But ferns can also reproduce in other ways. If there is insufficient moisture, they can grow from an unfertilized egg in a gametophyte. Or the tip of a frond of an adult fern

*See the chapter on growth for more on cellular automata.

can sprout a baby fern. When the parent frond droops and touches the ground, it develops roots and becomes a new independent fern. Or the roots of existing plants can allow new plants to form.

Clearly ferns are extremely successful at surviving and spreading, even if they are rather more delicate than modern plants. But I mention ferns for their structure, not for their reproductive strategies. Perhaps because they are such early forms of plants, ferns show more patterns than almost any other type of plant. Their patterns have a very special kind of regularity that we have not seen yet: *self-similarity.*

A fern does not have leaves; it has fronds. Most fronds look rather like giant feathers. But when you look at a frond, you'll see that it is made up of lots of smaller fronds growing from the main stem. And each of the little frondlets is made from even smaller frondlets, growing from their central stems. And each of the smaller ones is made up of even smaller frondlets. This is what self-similarity means: the frond is made from smaller elements that look the same as the whole.

Ferns don't stop there, though. When a frond grows, it *unrolls* from the middle of the plant. As the main stem unrolls, the smaller, miniature frondlets unroll along the sides. And as they unroll, the even smaller frondlets along their sides unroll as well. Watching a new frond unfurl is like watching a thousand gymnasts do forward rolls in different directions at the same time—an impressive sight.

All plants show some level of self-similarity. For example, the branches of most trees branch into similar smaller branches that divide into smaller ones that divide into twigs. When you examine the growth of each plant, it becomes clear that the number of times branches divide and the directions in which they grow seem to follow a set of rules. Each type of plant has a slightly different set of growing rules. The rules define how the stems, branches, and leaves emerge, and so give each type of plant its own distinctive shape.

The growth rules are contained within the DNA of all plants. And the DNA uses a set of rules rather than a list of specific instructions for each branch or frond simply because it's easier. Remember that it's the job of evolution to shape plants. It is very easy for evolution to find an instruction that says, "Branch into three, and make the smaller branchlets look like the original," and keep using the same instruction over and over again. It is much harder for evolution to develop a list of thousands of instructions that say, "Branch 1 will divide into three," "Branch 2 will divide into three," . . . , "Branch 736 will divide into three," and so on. It is also much more wasteful to use DNA in this way, and, as we know, nature tends to be efficient.

So patterns of self-similarity are caused by genetic rules being reused during the growth process. If the same rule is used over and over again for all parts of the plant, you end up with a very self-similar structure, like the frond of a fern.

Ferns display an exceptional degree of self-similarity compared to modern plants. The reason for this may be that the rules that ferns use for growth are still very simple. Our modern plants are quite adaptive. If there is more light or water in one area, or if one part of the plant is damaged, different rules may be used to change the number of branches and leaves to take advantage. Use many different rules, and the amount of self-similarity is reduced. The program of growth for modern plants may be slightly longer than the simple program used for ferns. But even in our cleverest plants, repetition of structure and self-similarity are clearly visible.

So it seems that nature loves patterns. Genetic rules, chemicals, golden numbers, bubbles, and many other clever tricks are all used to create the patterns so obvious in plants. But patterns are also very obvious inside the digital universes of our computers. Perhaps surprisingly, natural patterns caused by the golden number and the growth rules of plants are very important to computer scientists. So important that some researchers have become "digital gardeners," growing digital plants within their computers. Like biological plants, the digital plants follow their own rules of growth. And when we make the digital rules resemble the natural rules, the appearance of digital plants becomes indistinguishable from that of natural plants.

DIGITAL PATTERNS

We are in a large, dark room, 12 meters long and 6 meters wide. In front of us are five wooden columns, each with a different plant on the top. They are arranged in a semicircle around a large 4 × 3 meter screen. The screen is acting like the mirror in *Through the Looking-Glass,* showing us another room filled with plants. But time seems to be moving more quickly there. The plants in the other room are visibly growing before our eyes.

You approach one of the plants on its wooden pedestal. As you reach out to touch it, you suddenly notice on the screen that an identical plant is now growing in the other room. You see new stems push upward, leaves form and grow, with foliage pushing through the existing plants toward the light above. As your hands approach the plant, you see the growth in front of you change: the plant grows taller and more angled. You quickly discover that you can change the shape and type of the growing plants on the screen by moving toward and touching the plants in front of you. After some experimentation, you discover that you can make mosses, ferns, vines, and trees grow before your eyes. You also discover that with sufficient hand waving you can introduce a "killer plant" that clears space by removing parts of the lush vegetation before you.

A strange room, indeed. But despite the fun of creating digital plants, this is not just a toy. You are standing at a junction between two universes: physical and digital. The plants in one help create the plants in the other.

At this point you're probably wondering if I have led you into some make-believe place that I have conjured up out of my imagination. In fact, the room exists exactly as described. It is an exhibition in the ZKM Media Museum in Karlsruhe, Germany. The creation of artists Christa Sommerer and Laurent Mignonneau, this work demonstrates, in one of the most graphic ways possible, how we can grow digital plants that resemble biological ones. Their interactive computer installation uses biological plants as the interface to digital plants. Their computer senses how close your hands are to the plants on the wooden columns and grows digital plants in response. As you move, the computer adjusts the rules of growth, and the digital plants change dynamically.

This is an example of digital plants being used as art, but it is certainly not the only one. Artists have been growing digital forms using growth rules for many years. Jane Prophet, Paul Brown, William Latham, Alan Dorin, Rod Berry, and other artists all use plantlike growth to generate their art, with great international success. Many of them also use evolution to fine-tune the growth rules. They have shown that with the right rules, the resulting forms are as beautiful and original as the patterns in nature.

With the right rules. But what kinds of rules do we need in order to grow digital plants? And what should they do?

Not surprisingly, these are hard questions that have taken many years of careful investigation to answer. But by now you can surely guess where we found the answers: in the growth of biological plants. Earlier, we saw how the self-similarity of ferns is caused by genetic rules. Well, the rules that govern the growth of ferns, and indeed the growth of many other plants, are now quite well known. We may not have decoded the DNA that specifies the rules, but the study of plants and their growth has enabled us to interpolate many instructions responsible for constructing the form of plants.

The work of biologist Astrid Lindenmayer was perhaps the most influential in this area. Lindenmayer was interested in formalizing concepts of development in biology. Her idea was to create a *grammar* capable of describing growth. The grammar became known as a Lindenmayer system, or L-system.

L-SYSTEMS

At first glance, the idea that a grammar might have anything to do with the growth of plants may seem a little far-fetched. Most of us are used to the idea of grammar for language, not growth. If we're being taught English as a second language, for example, we learn that the sentences *The man sat on a chair* and *A chair sat on the man* mean very different things. The order of words indicates the relative functions of the phrases *the man* and *a chair* with respect to the verb. So we understand that the second sentence means the chair is sitting on the man even though chairs cannot sit and men can.

We also learn rules such as *The plural of fly is flies* and *The plural of deny is denies*. And we learn that there are always exceptions: *The plural of enjoy is enjoys* and *The plural of buy is buys*.

These complicated rules, full of exceptions and contradictions, are the things that our languages are all made from. They all have to be learned (or learnt in England), and until we know them, we cannot coherent: sentence Under-Standings, make—other people's from to, if you see what I mean. The grammar rules tell us how to construct valid sentences that others will understand.

But from a scientific point of view, such grammars are too informal. Their exceptions, ambiguities, and redundancies make them very useful for flexible (and vague) communication, but not so useful for the creation of structures with regular patterns. So, in the 1950s, linguist Noam Chomsky created formal generative grammars.

There are many types of formal grammars, but just to give you a sense, let's invent a simple one: a so-called type 2 context-free grammar. We start defining a formal grammar by specifying which words are allowed and how they can be arranged. For example, we might permit the following words to exist: *is, a, this, that, grammar,* and *Chomsky.* We also need to define different types and categories of words. In this example we can allow these: sentence, article, noun, proper-noun, noun-phrase, verb, and verb-phrase.

Finally, we need the grammar rules, to say how we should construct sentences using the words and categories. These are normally written in the form: $\alpha \rightarrow \beta$, which means "replace α with β." Let's provide some grammar rules for the example:

sentence → noun-phrase verb-phrase

verb-phrase → verb noun-phrase

noun-phrase → article noun

noun-phrase → "this"

noun-phrase → "that"

noun → proper-noun noun

proper-noun → "Chomsky"

noun → "grammar"

verb → "is"

article → "a"

So we now have a formal grammar with six words, seven categories, and ten rules. And it's all rather easy to use it to say something. We start off with the starting symbol "sentence" (also known as the distinguished symbol, hypothesis, or root). We then look at the rules and see that .we should replace "sentence" with "noun-phrase verb-phrase." We then see that "noun-phrase" can be replaced with "this" and "verb-phrase" can be replaced with "verb noun-phrase." "Verb" can be replaced with "is," while "noun-phrase" can be replaced with "article noun," which can be replaced with "a" and "proper-noun noun," which can be replaced with "Chomsky grammar":

sentence

noun-phrase verb-phrase

"this" verb noun-phrase

"this is" article noun

"this is a" proper-noun noun

"this is a Chomsky grammar"

The result is the sentence "This is a Chomsky grammar." You'll also find that other sentences are possible, for example: "A Chomsky grammar is a grammar," "A Chomsky grammar is this," and "This is that."

L-systems are even easier. They use the same idea of string-rewriting rules, but instead of words and categories of words, L-systems use symbols that represent biological features such as stems and leaves. To illustrate, think about another example. Instead of all those words and categories, we're going to have just two main symbols: Trunk and Stem. Instead of all of those complicated rules, we're just going to use one:

Stem → Trunk [left Stem] right Stem

As before, the rule means "Replace every 'Stem' with 'Trunk [left Stem] right Stem.'" But there are some new things in this rule. The "left" symbol means "angled to the left," the "right" symbol means "angled to the right," and the "]" symbol means "the next bit grows from the same place as the bit before the corresponding '[' symbol."*

Now we start things rolling by providing a stem and iteratively applying the rule:

---

*Or more precisely, "Return to the position and direction at the corresponding '[' symbol."

Stem
1. Trunk [left Stem] right Stem
2. Trunk [left Trunk [left Stem] right Stem] right Trunk [left Stem] right Stem
3. Trunk [left Trunk [left Trunk [left Stem] right Stem] right Trunk [left Stem] right Stem] right Trunk [left Trunk [left Stem] right Stem] right Trunk [left Stem] right Stem

After only three iterations of applying the same rule, the L-system sentence is now impressively long. It will continue to expand for as long as we keep applying the string-rewriting rule. And what it describes is the growth of a digital tree.

We began with "Stem," a single stem. After applying the rule, it became "Trunk [left Stem] right Stem," a tree with a trunk branching into two stems growing left and right. After the next iteration, it became a tree with a trunk branching into two, with two stems from each of the two branches. After the next iteration, it now had two stems added to the ends of the previous stems, resulting in a tree that now branched into eight. We could go on, but I'm sure you get the idea.

In a nutshell, this is how L-systems work. The same rules are applied iteratively, adding more and more complexity by rewriting and replacing elements. For more complicated digital plants, we use more complicated rules and more types of symbol. To add more realism, the rules can be slightly randomized, and aspects such as stem length and thickness can be added. But even with the basic kind of L-system we have looked at here, it is possible to grow digital plants with the same structure and patterns that biological plants display.

"But hang on," you may be thinking, "these are complicated strings, not plants." Quite right. The strings are the digital equivalent to the chain of protein activations and cell divisions caused by genetic rules in biological plants. Our digital plants are made by following the instructions provided by the L-systems. Digital trunks, stems, leaves, and flowers are created and placed according to the L-system growth program. The results can be displayed and really do look exactly like biological plants.

This is no coincidence. Some scientists have dedicated their careers to the understanding of growth and patterns in biological plants by finding the corresponding L-system rules. Przemyslaw Prusinkiewicz* of the University of Calgary is perhaps the most notable among them. Prusinkiewicz has shown just how closely the patterns in nature can be matched by different types of L-systems. His digital lilac, cypress, lily-of-the-valley, shepherd's purse, rose campion, and ferns, to name a tiny fraction, are all remarkable. They also show how complicated—and more often how simple—the growth rules need to be in order for these complex patterns to emerge.

L-systems are not only tools for examining biological patterns. There are a number of special applications that demand photorealistic digital plants. We've heard about the artists who use them to create their work, but the other main application is about as far from the art world as you can imagine: military pilot training.

Pilots require a huge amount of flight training in many different battle scenarios before they can be trusted to fly their multimillion-dollar aircraft in enemy airspace. They must be taught to make landings in tiny forest clearings. They must fly at hundreds of miles an hour just a few feet above the ground to avoid enemy radar. They need to twist and turn in the air to avoid missiles. So like all pilots, they spend many hours in flight simulators.

And what is a flight simulator? It is a digital universe in which pilots fly, fight, and crash without harm to themselves or any physical aircraft. As you might expect, realism in these digital universes is vital. The pilots must be fooled into thinking they're flying in our physical universe, and when flying very low, pilots use visual clues such as trees and bushes to judge height. Consequently, an important part of any flight simulator is the terrain and vegetation. So having digital trees and bushes that look exactly like biological ones is important. L-systems provide the solution in a number of today's better flight simulators. (Plate 11 illustrates just how realistic these trees can be.)

Art, flight simulators, and also . . . music. If you are musical, you know that all melodies have structures, repeating elements, and symmetries. In a mind-bending kind of way, this makes plants and music very similar. L-systems are grammars designed to represent the struc-

---

*Affectionately known as "Professor P" by the researchers in the field whom I spoke to.

ture of natural forms. It turns out that they can represent the form of music just as easily as plants. Instead of trunks, stems, and leaves, the symbols can be used to specify musical notes for different instruments. So many researchers have used L-systems to develop novel musical compositions, often with intriguing results. And it doesn't end there. If an L-system can define, say, a fern, and the same L-system can define a melody, then we can convert plant forms into music forms. We can *hear the shape* of ferns, trees, and shells. Which is exactly what I'm experiencing now. As I type, I am listening to the sound of a bush produced by programmer David Sharp's LMUSe software. It sounds surprisingly melodic, with staccato bursts of dramatic-sounding notes every so often. I'm now listening to various types of snowflake. This one sounds like very upbeat, contemporary dance. And here's a tree that sounds like jazz.

But enough of music. Finally, remember the Fibonacci series of 1, 1, 2, 3, 5, 8, 13, 21, 34, 55, . . . and how it provides the doorway to the golden section. L-systems also work in the same way. To make this clear, here's another L-system. We'll have two rules:

$$A \to AO$$

$$O \to A$$

where A stands for adult rabbit pair and O stands for offspring rabbit pair. The first rule says, "If you had an adult rabbit pair, now you have an adult rabbit pair and a pair of offspring." The second says, "If you had a pair of offspring, now you have a pair of adults." When we begin with a single newborn rabbit pair and iteratively apply the rules:

O
A
AO
AOA
AOAAO
AOAAOAOA

we see the Fibonacci sequence appear in the growth of the rabbit population, just as before. Replace rabbits with stems and trunks, and we have digital plants again. Replace them with musical notes, and we have melodies. They all divide and grow according to the same Fibonacci sequences. So L-systems make use of the golden number and other natural relationships as observed in nature. The patterns of digital plants really do follow the same processes as used in the patterns of biological plants.

L-systems are clever things. Like the natural patterns in ferns, they often produce more than is visible to the naked eye. But sometimes their patterns are endless—no matter how closely you look, there are always more patterns to see. We call these strange beasts *fractals*.

FRACTALS

You are probably already used to the ideas of fractals; we see things around us that resemble them all the time. In satellite photos, we see the jagged coastlines of continents. Zoom in, and we see more jagged edges, the size of towns. Zoom in again, and we see more jagged edges, the size of houses. Zoom in again, and there are more jagged edges, the size of people. And so on. There is the same kind of detail at all scales. But there are always limits. If we were to continue magnifying anything for long enough, we would eventually reach subatomic particles, where there is nothing but points of energy.

We looked at the self-similarity of ferns earlier—how each frond is made from smaller versions of itself, and how those frondlets are made from smaller versions and so on. But just like the coastlines of continents, the self-similarity of biological ferns is not infinite. As we continue to magnify a frond, we will eventually reach the scale of cells, and a new cellular pattern will emerge.

Digital ferns have no such limit. If we keep applying the same L-system rules, the self-similarity will keep on going. We could magnify the frond of a digital fern a million times, a billion times, or indeed forever, and we would still see new frondlets that looked like the original. This is a fractal fern (see plate 12).

So "never-ending fractals" don't really exist in our physical universe. But in the digital universe of computers, we can make them exist. We can have shapes with patterns that go on forever. Many, like the fractal fern, resemble the forms of plants. But some are unique.

Although L-systems (and related methods called *affine transformations*) can create fractals, the first and most famous fractal was discovered by the mathematician Benoit Mandelbrot in 1978. It is known as the Mandelbrot set, and some regard it as so important that they call it "the true geometry of nature."

Mandelbrot used a computer to visualize a very simple equation:

$$x_t = x_t^2 + c$$

It's simple because we have only two parameters here: x_t and c. If you're unfamiliar with math, just think of them as two boxes that you can keep numbers in. Our c box might hold the number 1, for example, and our x_t box might hold the number 3.

The equation means "Make the current value of x_t equal to the previous value multiplied by itself, plus the value of c." So if the previous value of x_t is 3 and the value of c is 1, then the current value is 3 × 3 + 1, which is 10. (You replace the number 3 inside your x_t box with the number 10.) And now, if the previous value is 10, then the current value becomes 10 × 10 + 1, which is 21.

Rather like the L-system rules, this equation is applied iteratively, transforming the current state into a new state. Instead of adding new branches and stems, it changes the value of x_t.

Things get a little more interesting, for Mandelbrot was interested in what this equation did for some unusual kinds of numbers. Not real numbers, like 3, 10, and 21. *Imaginary* numbers.

Imaginary numbers allow us to answer a hitherto impossible question: What is the square root of −1? Or to put it another way, what number can you multiply by itself in order to get −1?

The answer is not −1, for −1 multiplied by −1 is 1. It's not 1, for 1 multiplied by 1 is 1. In fact, it's not any real number. Find your calculator, type in 1, change the sign to −1, and press the square root $\sqrt{}$ button. You'll get an error.

The answer is i. If you multiply i by itself, you get −1. So i is an imaginary number. In general, these numbers are written as $n + mi$, where n and m are real numbers that tell how big the imaginary number is. So the length of $1 + 2i$ is less than $5 + 4i$.

Benoit Mandelbrot was interested in growth. Not the growth of plants, but the growth of values returned by the simple equation we saw earlier. He knew that for some values of c, the imaginary values returned by the equation would keep on growing. But for others, the values would oscillate between much smaller imaginary numbers. So he used his computer to draw a dot on the screen for each value of c that stopped the imaginary values from growing forever.* The resulting image was the Mandelbrot set—an inkblot resembling a squashed bug with little tendrils at the edges. Instead of being a pure geometric form such as a square or circle, it looked distinctly organic.

Its appearance was so strange that Mandelbrot tried magnifying the image—only to find more detail. The more he zoomed in, the more intricate structure he saw. Some of the patterns resembled distorted versions of the whole squashed bug, some resembled spirals, ribs, petals, and other biological forms. It soon became clear that the detail in the Mandelbrot set went on forever. You can magnify it for eternity and still see new shapes and structures emerge (see plate 13).

Today, our understanding of this fractal and others like it is much greater. Fractal art is commonplace, and researchers have also spent years studying the strange properties of the Mandelbrot set. Indeed, in 1991 a researcher named Dave Boll discovered quite by accident that the value of pi seems to be embedded in the structure of the fractal. Not only that, but our golden number phi is also evident in the spirals of the fractal. These fundamental constants that we discovered by looking at the natural forms around us are also emerging in the digital forms of fractals.

---

*In more detail: Mandelbrot wanted to know which values of c would make the length of the imaginary number stored in x_t stop growing when the equation was applied for an infinite number of times. He discovered that if the length ever went above 2, then it was unbounded—it would grow forever. But for the right imaginary values of c, sometimes the result would simply oscillate between different lengths less than 2.

Mandelbrot used his computer to apply the equation many times for different values of c. For each value of c, the computer would stop early if the length of the imaginary number in x_t was 2 or more. If the computer hadn't stopped early for that value of c, a black dot was drawn. The dot was placed at coordinate (m, n) using the numbers from the value of c: $(m + ni)$ where m was varied from −2.4 to 1.34 and n was varied from 1.4 to −1.4, to fill the computer screen.

Like L-systems, Mandelbrot fractals and their cousins can be converted into music. And there are other, more useful applications. One notable one that seems to be reaching maturity is image manipulation.

With the advent of multimedia, the Internet, and digital television, the need to push large images down slow telephone lines has never been greater. Already many formats are used to compress images and make them quicker to transmit. But fractal compression is beginning to look as though it might be another viable alternative. As we've seen, fractals use small, simple rules or equations to generate enormous complexity in images. Imagine if we could find a fractal equation for any image. We would then need only to transmit the equation, and the computer at the receiving end could calculate the image.

It's a very hard problem to find the right rules and equations for any image, but researchers have now made sufficient progress that fractals are being used for image compression.* Researchers report compression ratios of 40:1 or better for photographic images, with improved quality compared to the traditional methods such as JPEG. Although early attempts have not yet caught on, fractal decompression allows some unusual effects. Because the images are decoded from equations, they can be expanded to much greater resolutions (dots per inch) than the originals, with the equations filling in new detail. In a similar way, researchers suggest that fractal enhancement of images can be much better than standard ways of interpolating new pixels.

But this is where we must leave digital plants and their applications for now. We have seen how our digital universes have become colonized by L-system plants that have the same patterns as those in nature. We have also seen the native vegetation of digital universes, in the form of fractals. But there are other places where the patterns of fractal geometry have been discovered—not in Mandelbrot sets or plant growth but in patterns of change.

---

*Typically, images are divided up into smaller segments that appear similar to each other, and fractals are used to store one generic feature plus the transforms necessary to generate the other similar features from it.

TREE BALANCING

Walk in any woodland or forest, and you enter a battlefield. Plants are at war with each other. The pine trees are battling with the oak trees, the ferns are fighting the bluebells. You may hear only the rustle of leaves and songs of birds, but the war ranges around you.

The battle is being fought over resources: the sunlight, water, and soil that every plant needs to survive. For every tree you see in the forest, hundreds died trying to grow in the same soil and fill the same patch of sunshine. Plants fight over territory more fiercely than we do in any of our wars.

Because of limited space and immense competition for resources, plants (and indeed all other forms of life) must compromise. A delicate balance of different species always emerges, dynamically changing and oscillating as different species gain more territory but are then fought back. Often species will form codependencies, with the ground-hugging shrubs relying on the dappled light and moist conditions provided by the trees above. A typical forest forms an unimaginably complex network of competition and codependencies, the ecosystem balancing on a tightrope between order and chaos.

After the previous chapter, this may be sounding a little familiar. We're looking at another example of self-organization, but this time, instead of multiple agents of the same species working together, we have members of different species that are forced to coexist. To see an example of the strange patterns of change that these relationships cause in ecosystems, let me take you to Kangaroo Island, in southern Australia.

Eucalyptus trees are indigenous to Australia and Tasmania. They have fragrant, oil-rich leaves that many species of insects and animals find irresistible. On Kangaroo Island, as you might imagine, there were a large number of kangaroos, many eucalyptus trees, and a delicate balance between the populations of all the various species that made up the island ecosystem. But in the 1920s, humans introduced the koala bear to the island.

Koalas love eucalyptus trees; it is their main source of food. When a small number found themselves on Kangaroo Island, they discovered food everywhere. Not surprisingly, they flourished, becoming exceedingly numerous. But over the subsequent years, their numbers grew too large. At the time of writing, the number of koalas is so high that the eucalyptus trees are being overgrazed and are dying. So many trees are dying that large areas of soil are being left bare, where they are eroded by wind and rain, causing large changes to the habitat. So many are dead that there is now insufficient food for all of the koalas, and so, unless someone intervenes, thousands will die of starvation.*

For now, let's assume that nobody does intervene. As we move forward in time, the koala population will fall dramatically. But as the number of koalas eating eucalyptus trees is reduced, the trees will start to recover. Before long, we will find that there are far more eucalyptus trees and fewer koalas. So with more food, the population of koalas will increase again. This will reduce the number of trees, which will reduce the number of koalas. And so on. The two populations are locked together in an oscillating pattern: when one goes up, the other goes down, and vice versa.

Recent research has shown that these patterns can become quite localized. Different areas may have different patterns of population oscillations. And (through another mind-bending connection) it seems that these different oscillating patterns of population sizes behave in the same way as the reaction-diffusion chemical patterns (which may create designs on shells) that we saw earlier in this chapter. The chemical patterns are caused by different chemicals enhancing and suppressing each other. The population patterns are caused by different populations enhancing and suppressing each other.† Nature seems to reuse the same ideas for many different things.

But I digress. There's something a little strange about our oscillating pattern of eucalyptus tree and koala populations. It is completely deterministic—we can work out exactly how many trees it takes to feed how

---

*Kangaroo Island is managed by humans, so it rather looks as though there will be intervention in this case. Polls suggest that Australians do not like the idea of koalas dying, whether by natural causes or by culling, so the current ideas seem to favor reducing their fertility and/or relocating them, to enable the eucalyptus trees to recover.

†And the same computer technique used by biologists to model chemical pattern formation—cellular automata—is also used by ecologists to show spatial population pattern formation.

many koalas and how the populations will change in the short term. However, we can't predict anything in the long term. The pattern may be deterministic, but it is *chaotic*—it is very sensitive to minute changes over time. So if a single eucalyptus tree manages to live just another day, it will imperceptibly change the oscillating pattern in the short term, but may cause huge changes to the population numbers in the long term. Fluctuations are amplified, making these chaotic patterns unpredictable, despite their apparent simplicity.

Unpredictable chaos. Somehow it doesn't sound as though we can do much with such things. Indeed, that's what most people thought, until a new way of looking at such systems was introduced. Chaotic patterns may be unpredictable, but that doesn't mean we can't use them.

DIGITAL DYNAMICS

The breakthrough came when chaos and computers came together. A computer was used to draw the oscillating patterns of population changes, plotting the size of each population as x and y coordinates. The twin up-and-down oscillations were then transformed into ellipses. What we would see for our tree and koala populations would be a series of ellipses, each one not quite on top of any other. Every ellipse would be subtly different, illustrating the subtly different numbers in the two populations each time. But each ellipse would also follow the same general path, loosely orbiting some point in the middle.

These displays are types of fractals. No matter how closely you look, there will always be more ellipses, and yet none will ever quite overlap. They show how regular the population oscillations are, and yet how difficult it is to work out exactly what the precise numbers in each population will be in the future. But most of all, they show a *strange attractor*—the general ellipsoid shape that governs the overall movement of the system. If we can see the strange attractor, then we can gain understanding about the dynamics of chaotic systems. In this example, we can see that the population oscillations will always continue and will always form a certain shape. In other examples, such as the flow of water, the prediction of weather, or the normal electrical activity of brains, the strange attractors are more complicated shapes (see plate 14). But by knowing the shape of a strange attractor, we are able to gain understandings of normal behavior and even gain some control of the chaotic system.

Our computers allow us to do this. One of the first applications of chaos was in the control of satellite ISEE-3, the third International Sun-Earth Explorer, by NASA. This old satellite had almost used up its supply of fuel, but still held useful instrumentation that would allow it to make some invaluable observations of a nearby comet. So NASA applied chaos theory to the orbits of the Sun, Earth, and Moon and determined exactly when to squirt its thruster. The minute changes in orbit were then amplified through the effects of chaos, enabling it to swing around the Moon five times and fly off into space, right through the tail of the comet. The satellite, now renamed ICE, the International Cometary Explorer, was the first to make direct measurements within a comet's tail.

The application of chaos theory has also taken place in consumer products. Goldstar Co. says it was the first in this arena with its "chaotic washing machine," which was created in 1993. Goldstar added a small pulsator to stir the water in addition to the stirring caused by the main rotation of the drum. The result—it claims—is cleaner, less tangled clothes.

Chaos theory modeled within our computers also permits more significant results. Some of the most exciting are the chaos control systems that have made erratically beating hearts return to regular rhythms and methods that may lead to the prevention of epileptic seizures.

Even chaos, it seems, can be tamed within our digital universes. From the patterns of changing tree populations to the control of spacecraft, when a natural process is translated into our computers, our technology makes several giant leaps forward.

SUMMARY

So that's plants and patterns for you. Diatoms, pine cones, snail shells, and ferns show how bubbles, golden numbers, chemicals, and genetic rules are all used to create the patterns so obvious in plants. Special grammars known as L-systems seem to work in the same way, using growth rules and golden numbers when they generate their patterns within computers. But the native digital fauna, fractals, are just as interesting, with their strange forms and patterns that go on forever. And fractals are not unlike the mysterious dynamics of population changes in ecologies. These chaotic patterns are found everywhere in nature. They may be hard to predict precisely, but they can be understood and used within our computers, with some impressive results.

The patterns we looked at here are everywhere, of course—not just in plants but in every form of life. And there are some patterns and organizations within your body right now, keeping you alive. They work together, a society fighting for your life every second of your existence. I am talking about your immune system.

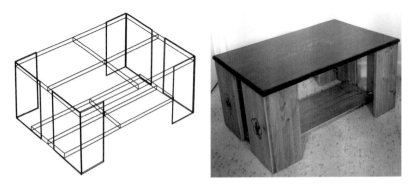

1. Evolving coffee tables. The image on the left shows the design evolved by a genetic algorithm, and the image on the right shows a photograph of the coffee table built to this design. (Copyright © 1998 Peter J. Bentley)

2. *Hyperspace Embryo,* an evolved fractal image using genetic programming. (Copyright © 1999 Steven Rooke)

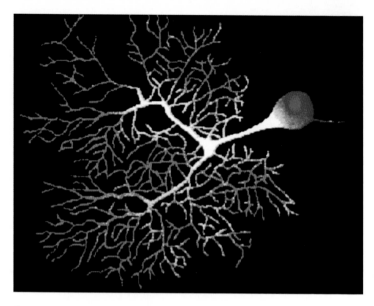

3. A neuron in the cerebellum as it fires. The dendrites are shown in red, the cell body in blue. A small part of the axon can be seen at the right of the picture. (Copyright © 1996 Bell Labs/Lucent Technologies, all rights reserved. Reprinted with permission)

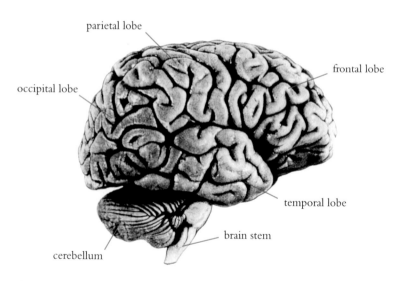

4. The human brain. (Photograph courtesy Professor S. Zeki, copyright © Laboratory of Neurobiology, UCL)

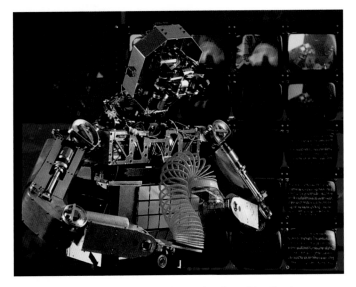

5. Cog playing with Slinky. Cog was developed by Rodney Brooks and the Humanoid Robotics Group at MIT's AI Lab. (Photograph copyright © 2000 Peter Menzel Photography)

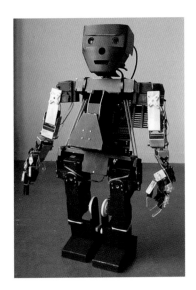
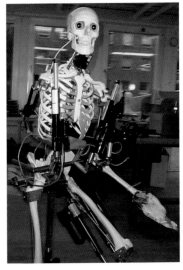

6. Sixty-centimeter-tall Elvis and his unborn sister, Priscilla. These humanoid robots have brains that operate in a similar way to Brooks's subsumption architecture, except that each layer of their brains is evolved, not designed by humans. (Copyright © 2001 Peter Nordin, coordinator of the Humanoid Project, Goteborg University)

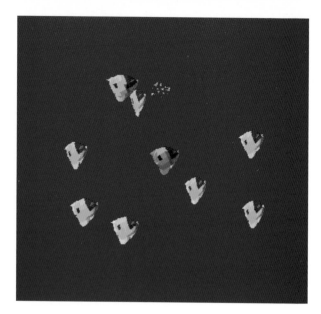

7. Schooling tropical fish eating food. A screen shot from a Java applet by Bill Kraus at www.digitalbiology.com using the boids algorithm originally developed by Craig Reynolds.

8. Diatom "art" (*left*) and computer models of diatoms to be used for virtual crash tests (*right*). (Copyright © Alfred Wegener Institute for Polar and Marine Research, Germany, and copyright © 2000 Christian Hamm, AWI, respectively)

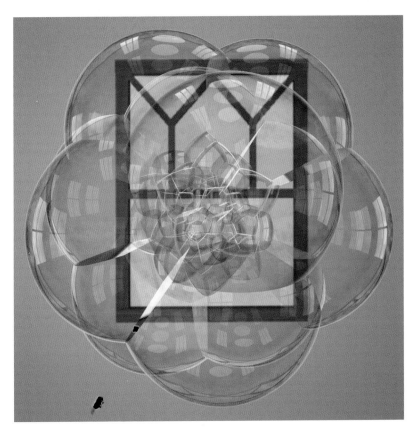

9. *120-Cell Soap Bubble* by John Sullivan (computer–generated image).
(Copyright © 1991 The Geometry Center, University of Minnesota. All
rights reserved)

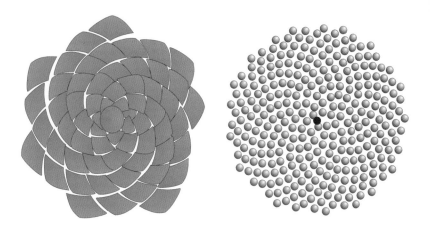

10. Leaflets of a pine cone (*left*) follow the same golden ratio as seeds in a flower seed head or meatballs on a plate (*right*). (Copyright © 2001 Peter J. Bentley)

11. What's natural and what's digital? A thirty- by forty-inch LightJet print from "The Landscape Room" by Jane Prophet, featuring fractal trees by Gordon Selley. (Copyright © 2000 Jane Prophet and Gordon Selley)

12. A fractal fern with never-ending self-similarity. Generated by Fractint v18.21.

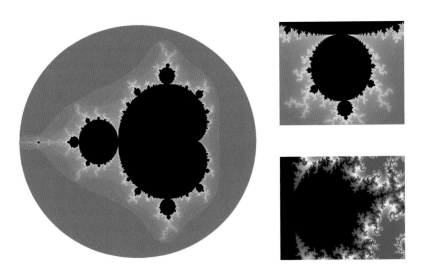

13. The Mandelbrot set. The image at left shows the entire set; right top shows a magnification of lower area; right bottom shows a further magnification. Generated by Fractint v18.21.

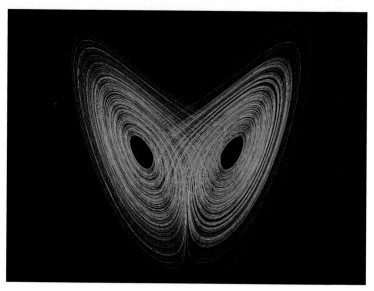

14. Lorenz's strange attractor, describing the chaotic motion of a water-wheel. Generated by Fractint v18.21.

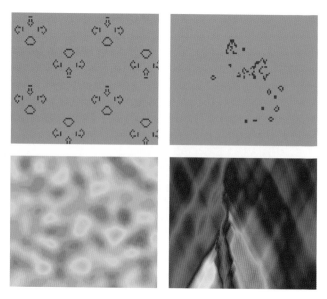

15. Images developed using cellular automata: *Conway's Life* (*top left and right*), *Leopard skin* (*bottom left*), and *Boiling Wave* (*bottom right*). Generated by CAPOW! Cellular Automata for Power Grid Simulation, Electric Power Research Institute, Palo Alto, CA.

7

IMMUNE SYSTEMS

WITH A SOUNDLESS BURST, you are released from a gigantic womb into the surrounding fluid. You float, incapable of moving yourself, with your hundreds of siblings all around. But there is movement. The fluid that supports you is awash with activity—vast unknowable things that move, searching and eating. You are in mortal danger, born into a hostile world where everything exists to kill you.

Suddenly the whole universe moves. First, everything is sucked backward. Then there is an explosion of motion forward. The warm, comfortable fluid that supports you breaks up into vast ocean-sized globules. You and millions of your relatives fly through the air in a spaceship made from a pinhead-sized drop of fluid.

In less than three seconds, your epic flight from peril is over. The fluid-spaceship splashes down onto a warm wet surface, along with a rain of other similar drops. You are thrown onto a dark wet wall, which seems to stretch forever in all directions. But you are fine. Your intricate spherical shell made from a mosaic of interlocked protein molecules protects the special cargo held inside.

For an unknowable amount of time, you drift on the thin liquid coating of the wall. Eventually you bump into one of the building blocks of the wall—a gigantic mass one thousand times your size. You are dwarfed and insignificant, smaller than a grain of sand on a beach ball. And you are stuck. Your shell has become caught in the protuberances of this monstrous creature. Even worse, the sides of the creature now begin to move around you. Within moments you are engulfed, your shell ripped open.

But all is not lost. Even though your shell is no more, you are still unharmed, within the belly of the beast. And it is only now that you discover that this is exactly where you want to be.

The monster that swallowed you is a machine. From inside, it is apparent that there are vast factories, making protein molecules that swim around you, joining together and breaking into new shapes according to some grand master plan. And your presence is causing a disturbance. Proteins and other molecules are becoming attracted to you. They soon begin to follow your instructions, fooled into obeying you. Before long, you have the whole vast machinery of your prison working for you alone. The space becomes filled with new molecules—long strings of RNA encased within protein shells—as you instruct the factories to make identical copies of yourself. You have no interest in preserving the monster that confines you, so you pay no attention to its chemical cries of exhaustion. Eventually, it can work no more and bursts, releasing hundreds of your children as it does so.

As it dies, so do you, for you find yourself flushed down a vast pipeway into a huge sea of acid where the RNA that you are made from is quickly destroyed. But your children live on. They spread across the giant wet wall, sticking to and infiltrating it. Every few minutes, more of the vast cells burst, releasing tens of thousands more of your children. Soon the giant wall has many spots of infection all over it, and it begins to run with fluid as other cells are stimulated to generate and release mucus.

Within the huge cells, your children continue to use the factories to duplicate themselves. As they do, some of the protein molecules created during the manufacturing process get stuck to other protein molecules, still being created by the cell. These push their way through the cell wall like seedlings through soil, exposing samples of your children's proteins to the outside world. And there are predators out there waiting for exactly these signals. Huge, colorless cells, very different from the pink mucus-producing cells that your children are reproducing in, begin to cluster around every infected cell. No mercy is shown as the infected cells are destroyed—and with them thousands of your unfinished children.

For some time, a battle rages. It is an arms race between the infection rate of your children and the colorless killer cells. As the number of killer cells increases, so a new chemical cocktail released by them increases.

The vast walls respond, running with more fluid and becoming swollen and red. Occasionally, spasms wrack the walls, forcing huge droplets of fluid out into the space beyond. Like your own flight from danger, these fluid spaceships carry millions of your children away. This is the last trick your offspring can play: fooling the killer cells into providing an escape route for some lucky progeny.

And now the war is nearly over. The colorless cells and their chemicals saturate the fluid, turning it a greeny-yellow color. All infected cells are destroyed. Even those children left floating alone outside the cells are not safe. New colorless cells have started to appear, emitting swarms of proteins that stick to the spherical shells that protect the children. Bound and gagged, they can no longer stick to anything, rendering them sterile and doomed to a short, futile life.

Eventually your children are all dead in this world. All that remains are the patrolling colorless cells, now trained to kill on sight anything that resembles you. They will not forget you.

STAYING ALIVE

This is what a common cold feels like from the perspective of the cold. What may be a nuisance two or three times a year for us is a life-and-death struggle for survival by a new cold virus.

In the story above, you played the role of a rhinovirus—a type of virus responsible for up to 70 percent of colds according to some estimates. Like all viruses, it is so primitive it cannot really be called alive. A collection of proteins surrounding a coiled string of RNA (or DNA for some other types), viruses are little more than highly complicated molecules. They are a thousand times smaller than most of our cells, and they cannot even reproduce by themselves. To spread, they rely on deception and mendacity. First, they fool individual cells lining the insides of our noses and throats to let them enter through the cell wall. They then trick each cell into following their viral RNA instructions instead of the cell's own DNA instructions. They even manage to ensure their own survival by causing the killer cells of our immune systems to trigger excessive mucus production and irritation. This results in coughs and sneezes that spray copies of the viruses onto new hosts.

But however clever they may be, our immune system is cleverer. We have armies of dedicated soldiers that never sleep within our bodies. They fight for our lives every second of every day of our existence. Without them, every time we are exposed to a new virus, bacterium, fungus, or parasite (which is happening at a rate of several every minute), we would become permanently and often terminally afflicted. With them, we recover. And every time our immune system encounters a new threat, it *learns*. Once it has fought off an attack, we will rarely suffer from the same virus, bacterium, fungus, or parasite again. Somehow, we develop an immunity, which protects us for the rest of our lives.

In this chapter we explore how these seemingly miraculous processes happen within us. We'll take a look at a few of the mechanisms that keep us alive. Although some of the details can be ferociously complicated, many of the techniques used by our immune system—things such as evolution, memory, and self-organization—will be remarkably familiar.

UNKNOWN SYSTEMS

Unlike such things as evolution and brains, immune systems do not get much coverage in the general literature. So while most of us are familiar with ideas of survival of the fittest and neurons, few really know anything about their own immune systems. The fact that the terminology of immunology is largely impenetrable doesn't help, nor does the fact that new theories about the working of the immune system are still emerging, causing controversy and confusion.

But the lack of information about such an important part of ourselves is surprising. For example, did you know that unlike our brains, our immune system is a distributed system, spread throughout the body? Or that it uses many different organs such as the pancreas, thymus, lymph nodes, and bone marrow to help make all of the many white blood cells and chemicals that protect us? Did you know, for instance, that your bone marrow (the soft core inside your bones) makes all the cells in your blood? Or that allergic reactions are caused by an overreaction in your immune system? Or that we actually have a second "plumbing system" used by the immune system in addition to blood vessels, called the lymphatic system? Well, now you know.

In fact, let's take a closer look at the lymphatic system right now.

LYMPHATIC SECURITY

In our bodies, we have blood vessels that carry nutrients and red blood cells (which store the oxygen you breathe in) to every cell in the body. Pumped by the heart, our blood sustains and maintains the other cells, keeping them alive. Our arteries feed blood through smaller and smaller branches, eventually letting it seep from tiny capillaries into the cells of the organs, before being sucked back through veins. Blood vessels often have muscles lining their walls, allowing them to contract or enlarge openings. This allows your brain to adjust blood flow to every part of your body—one reason that your hands feel cold when you are stressed and warm when you are relaxed. When stressed, your brain redirects blood to organs that might aid you should you need to fight or flee.

The trouble with this system of blood supply is that is leaks. Fluids and smaller substances are not all sucked back into your veins; some get left behind in your tissues. This is where the lymphatic system comes in. Wherever you have capillaries (and that's *everywhere*), you also have lymphatic vessels, which suck up all the spillage and return this fluid into your blood supply (at areas such as the jugular vein in your neck).

In mammals, the lymphatic system also plays an important role in the immune system. Because spilled fluid is drained from every area of your body, should any invading pathogen such as a bacterium or virus enter your body (for example, through a scrape on your knee), it will be sucked immediately into the lymphatic vessels. And evolution, being an efficient designer, has placed security barriers called lymph nodes throughout our lymphatic system in order to catch these invaders. We have lymph nodes deep within our chests, under our arms, in the groin region, and under our chins. When the doctor feels under our arms and chin, he is trying to feel whether our lymph nodes are swollen. If they are, the chances are that your immune system is busily fighting some kind of infection nearby.

UP THE ANTI

If our lymph nodes are the security barriers, how do they catch the harmful intruders? The work is done by a special team of different cells working together: macrophages, T cells, and B cells (together commonly referred to as white blood cells).

It may seem a little crude, but the first thing that happens is that large cells called macrophages simply eat everything they can, wherever they are in our bodies.* The word *macrophage* derives from the Greek meaning "big eater." And these cells are aptly named, for they spend their time eating everything that goes by. They eat our own dead cells; they eat invading bacteria and all other pathogens. They have large appetites.

But however much macrophages eat, this is not usually enough to prevent infection by pathogens. Indeed, many types of pathogens (such as viruses and parasites) reproduce inside a macrophage, using the cell's own internal machinery to help create copies. What the macrophages are actually doing is helping to investigate everything. A macrophage doesn't just eat things, it presents the crumbs of its meal on its outer surface. While it's doing this, it migrates to the security barriers of our lymph nodes. And there happens to be another kind of cell ready and waiting, which specializes in examining macrophage crumbs and working out what to do next. This is the job of the T cell.

T cells (like all immune system cells) cannot recognize bacteria, viruses, or even self-cells directly. Instead, they use *antigens* to tell them what everything is. Antigens are nature's bar codes—unique patterns of proteins that coat the surface of all organisms. Recognize the antigens, and you recognize the organism (or pathogen). And T cells have some of the best bar code readers around. So it's the job of the T cells to examine the macrophage crumbs (which are really antigens stripped from the surface of the digested organism, attached to special macrophage MHC molecules) and work out what to do.

Luckily, T cells are clever little critters. When examining the surface of a macrophage, they are able to work out whether the big eater has just eaten something it shouldn't—a pathogen. If they detect such a foreign substance, they swing into action.

There are actually two main types of T cells: the helper T cell and the killer T cell. Let's look at the helper T cell first, since it tends to coordinate most of the other important cells.

---

*It has recently been discovered that *dendritic cells* are perhaps most important for this task in our bodies.

If a helper T cell spots that a macrophage has eaten something nasty, it immediately looks for a B cell to report the problem to. The lymph nodes contain many immature B cells, fresh from the bone marrow, that wait for just such an emergency. Our helper T cell tells the B cells all about the problem by releasing a chemical message (lymphokines). This causes the B cells to mature and begin releasing proteins called *antibodies*. These antibodies are released into the surrounding lymph fluid, and quickly propagate into the bloodstream and throughout the rest of the body. The antibodies are highly specific to the antigen belonging to the pathogen that was eaten by our macrophage: if they encounter other pathogens that look the same, they will stick to them. It does not take long before antibodies have stuck to all copies of the invading pathogens. These will either kill the pathogens directly, or our macrophages will come back to play. To a macrophage, anything with an antibody on it is especially tasty, so these big eaters happily go around consuming all the pathogens that have antibodies stuck to them.

So our immune system removes pathogens by first using macrophages to examine (eat) some of them, then using helper T cells to activate B cells, which generate antibodies that stick to the pathogens, which are then eaten by macrophages. Complicated but clever.

But what if the pathogens are clever too? If the invader happens to be a virus, then it will hide inside our own cells. How are these pathogens removed?

This is where the killer T cells play their starring role. In the lymph nodes, when a helper T cell notices that a macrophage has eaten a virus, it will again cause B cells to make antibodies for that virus, but it will also help killer T cells to mature. An ever-increasing squad of helper and killer T cells then patrols the body, clustering around any cells infected by the virus. Egged on by the helper T cells, the killer cells kill any of our cells that have become infected. They also release chemical signals that encourage macrophages to cluster around the area, causing *inflammation*. For some infections, these chemicals may tell our brains to raise our body temperature. Many pathogens are very susceptible to temperature. Even a slight increase may destroy them or slow down their reproduction rate. Having a fever can be a good thing.

So even if you are a virus, capable of hiding inside cells, the immune system will find you. It will kill any cells you are inside, and it will attack you with antibodies and macrophages if you are outside in the bloodstream.

In essence, this is how our immune system keeps us alive. It maintains an ever-vigilant watch over us, using its own version of the emergency services. Instead of fire, ambulance, and police departments, our immune system has macrophages, B cells, and T cells. The "fire department" macrophages go everywhere and do the heavy work. The "ambulance" B cells dispense their medical aid in the form of antibodies. And the "police" T cells help the other departments and work to remove any antisocial pathogens.* They all act together in a highly coordinated and targeted manner to keep us healthy.

HOW TO BE A GOOD IMMUNE SYSTEM: DIVERSITY, SPECIFICITY, MEMORY, AND TOLERANCE

Of course, this is all a gross simplification of what is really going on. There are actually many types of macrophage, B cell, helper T cell, and killer T cell, as well as a hugely complex chemical communication system used to coordinate everything. But even if we ignore these details (which we will), things are still considerably more difficult for our immune systems. There are millions of different types of bacteria, virus, parasite, and fungus that can be harmful to us. There are also many types of bacteria that we positively need in order to survive—we would be unable to digest our food without the bacteria that live in our gut, for example. There are also many different types of cell in our own bodies. How can the cells of our immune system tell the difference between harmful pathogens and the cells and bacteria that we are actually made from? We know our immune system can do this and more—for example, when we transplant an organ from one person to another, the immune system detects the foreign organ and tries to reject it. How does it know that the cells of a new kidney do not belong there, when the cells of the old kidney performed exactly

---

*They even have their own version of bullets: special molecules that chemically punch holes in our infected cells (lysis) to destroy them.

the same function? We also know that the antibodies and other responses of the immune system are very specific to pathogens. How does it learn how to remove these brand-new pathogens so quickly? And once learned, how can the immune system remember the pathogen, thus preventing infection from the same invader again?

These are the hard questions in immunology. Although we don't have all of the answers, we know enough to get a broad view of how our immune system operates. And perhaps the four most important features of the immune system are *diversity, specificity, memory,* and *tolerance.* New pathogens are so successfully attacked by our antibodies because B cells are able to make an enormously *diverse* range of different antibodies—at least one for every possible pathogen. Each antibody is a specialist, carefully designed to attack and kill *specific* pathogens. Our immune system *remembers* pathogens because it ensures that some of the B cells that make the corresponding antibodies persist in our blood. And the antibodies (and other cells) do not attack our own self-cells because they are *tolerant* of self.

This may still seem rather mysterious, so let's look at how these abilities come about. After the chapter on evolution, the mechanisms should soon begin to feel a little familiar.

THE GOD QUESTION

The main job of the B cell is to generate antibodies. These special molecules are carefully designed to stick to any new pathogen encountered and will either kill the invader outright or ensure that it is eaten by our macrophages. The B cell has two problems: first, how to create antibodies that will stick to any possible pathogen and, second, how to create antibodies that do not stick to any self-cells (us). If the diversity of possible antibodies is insufficient, then at some point a pathogen will come along that our B cells cannot deal with and we will be defenseless against the disease. But if the B cell creates antibodies that stick to our own self-cells, then our macrophages will eat our cells. With our own immune system consuming us, we will quickly die.* A careful balance

---

*Unfortunately, some pathogens are able to fool the immune system into doing exactly this. Such illnesses are called *autoimmune diseases.* We'll look at some of these later.

between death by disease and death by immune system has to be maintained at all times.

B cells make their antibodies by following the instructions in the DNA held within them. Like almost all the cells in our bodies, B cells are like little machines, dedicated to following the DNA program written into them by evolution over thousands of millennia. (The chapter on growth explores exactly how this happens.) And so for many years, it was believed that the instructions needed to generate every possible useful antibody must be held within our DNA. For every one of the hundreds of millions of possible threats, there must be an instruction on how to make the corresponding antibody. But when we started to unravel the code of our DNA (through research efforts such as the Human Genome Project), we discovered that this could not be true. If instructions on how to make every possible, useful antibody were stored in our DNA, there would not be enough room left for the instructions needed to grow us. Clearly, the diversity of antibodies was being created inside B cells in a brand-new way. But how? This was the GOD (generation of diversity) question in immunology.

The solution that nature found will be familiar to anyone who likes candy shops, specifically the kind where you can fill a bag with any combination of candies from dozens of different kinds available. This allows you to vary your choice every time: twelve licorice, ten butterscotch, five cola bottles or four licorice, four cola bottles, five yellow gummies or ten chocolate peanuts, five fudges, ten toffees. You could go back every hour for the rest of your life and always get a different combination (not that your teeth would appreciate such a practice).

It turns out that B cells use the same "pick and mix" method for creating their special proteins. Through a convoluted and impressively designed gene expression method, B cells make new antibodies by taking random elements from a range of different possibilities. In the same way that our candy store allows us to have many millions of different candy combinations by picking different candies from the hundred or so available, B cells are able to make many millions of different antibody proteins by combining about 320 fragments in different combinations. This was an astonishing discovery, for it is absolutely unique in biology. Normally, cells generate whole proteins directly from evolved DNA instructions. To discover that B cells were actually *building* proteins from fragments was remarkable.

Remarkable but very sensible. Evolution helps design creatures so that they will be more likely to survive in a given environment. But pathogens such as viruses and bacteria reproduce and mutate so quickly that they evolve at a much faster rate than we do. This is why we repeatedly suffer from the common cold throughout our lives. We become immune to each cold virus we encounter, but the viruses evolve so quickly that we are presented with brand-new versions every few months. At this rate of change, there is no chance for evolution to provide our immune systems with all the answers in our DNA. Instead, evolution has given our immune system the ability to make educated guesses and be sufficiently adaptable to cope with new pathogens.

EVOLVING PRECISE CLONES

Diversity of antibodies by clever guesswork is one thing, but how is it that our B cells can be so precise? Whenever a new pathogen invades, it does not take long before antibodies tailored to the antigen (bar code) of that pathogen are being manufactured in large quantities by B cells. How does the immune system convert guesswork into specific, controlled antidotes?

To explain, let's think about music. The popular music industry is interested in only one thing: money. If it spots that a boy band is proving very popular or a pretty young, solo female artist is doing well, then it wants to cash in on that market. So each company finds or puts together its own clones. In the 1990s, there was a boy band cloning frenzy—huge numbers of bands constructed according to a set formula. In the early 2000s, it looks as though pretty, young female singers are being cloned by the dozen. By duplicating successful bands and discarding unsuccessful ones, the music industry happily makes large amounts of money.

Thankfully, unlike the music industry, your immune system does have your best interests at heart. But it's as successful at mass-producing the right formula (for new antibodies) as the music industry is at mass-producing formulaic new bands. Like the music industry, our immune systems discard unsuccessful B cells and clone successful ones.

They do this using a process known as clonal selection. Whenever B cells are activated (usually by helper T cells that have spotted that a macrophage has eaten something nasty), they begin to manufacture antibodies. Rather like a band releasing a single, each B cell makes one type of antibody and presents the protein all over its surface. Because of the huge diversity of antibodies produced, for every few thousand releases, one or two will normally turn out to work rather well. So when other copies of the antigen float by, they stick to the matching antibody on the surface of the B cell. And when the B cell discovers that its chemical music is popular, it clones itself. One becomes two, two become four, four become eight, and in a matter of a few hours, there is a large population of these clever B cells in the bloodstream (now differentiated into plasma cells), all producing the right kind of antibody. This saturates us with antibodies that match the antigen (bar code) belonging to the pathogen, ensuring that the most successful antibodies are used immediately. It's just like when the music industry saturates the market with cloned bands that all produce the same kind of successful music.

But this is where the current dubious analogy breaks down, for our immune systems are much cleverer than the music industry. B cell clones are not always identical. Let's forget about music and look at the biology in more detail.

As the B cells clone themselves, some undergo many genetic mutations. Known as somatic hypermutation, this is the immune system's way of keeping up with the fast evolution of viruses and bacteria—it uses evolution itself. All of the cloned B cells resemble their parents, but because of the mutations, some generate slightly different versions of the antibody. Those B cells that produce a better antibody for the current antigen will produce more offspring by cloning a little more often. Those B cells that produce a slightly worse antibody will not clone so often; the antigens are more likely to stick to the better antibodies. And as the generations of B cells go by, they become better and better at generating exactly the right antibody for the current antigen. Within our bodies, these cells evolve in a matter of days to tackle new diseases.*

---

*We can check that it is evolution by looking for the three processes: reproduction with inheritance, selection, and variation. The B cells reproduce with

It is a truly inspired scheme—use the pathogens' best weapon against them. If they evolve quickly, just make sure the immune system cells evolve quickly too. There's only one major drawback: because our B cells evolve inside our bodies quite independently of our own evolution, there is no way for the changes in DNA structure within the B cells to find their way back into our "germ-line" cells. Our sperm and eggs cannot have their DNA updated with the evolved DNA of our B cells, so we cannot pass on all of our immunities to our children.* Instead, evolution gives our immune system the capability to learn for itself. But this is not to say that we do not pass on many of our immunities to our children. The most important ways we do this are in the womb and through breast milk. Human milk is very high in macrophages and antibodies—one reason that breast-feeding is so important for infants during the first few months of their lives.

REMEMBERING TOLERANCE BY STAYING ALIVE

B cells use evolution to become specialists at generating antibodies for every new pathogen. But how do our immune systems remember these evolved immunities, keeping us from suffering from the same pathogens for the rest of our lives? Why doesn't our immune system simply lose these specialists as a result of the natural turnover of cells in our bodies?

It seems that we are designed too well for things like that to happen. When activated B cells are cloning themselves, instead of generating plasma cells to churn out antibodies, sometimes they generate *memory*

---

inheritance, for they clone themselves and pass copies of their DNA to their children as normal. There is selection because those B cells that make better antibodies are selected over those that make worse antibodies—the better ones have more offspring. And somatic hypermutation of the DNA within the B cells provides the variation. So the next time someone questions the existence of evolution, remind them that their B cells are evolving within their bodies as they speak.

*Ted Steele, a researcher of molecular immunology and evolution, still tries to argue that this may actually be possible with the help of special viruses called retroviruses, but his Lamarckian theory of soma-to-germ line transcription has never been more than an unsubstantiated hypothesis with little support in the field.

cells. These unusual cells do not immediately produce any antibodies. Instead, they are extremely long-lived cells that lie in wait—a built-in ambush squad. Even when all of the active B cells are long dead and you no longer have antibodies for a particular pathogen's antigen (bar code) in your blood, these memory cells persist. And if, some years later, you should encounter the same pathogen, the memory cells leap into action, rapidly cloning themselves and producing the appropriate antibodies before the pathogen has had time to look around. You may not even notice that you have been infected, the response from your immune system is so quick. You are immune to this illness, thanks to the excellent memory of your immune system.

So "pick and mix" proteins, clonal selection, and memory cells generate diversity, specificity, and memory. These abilities help our immune system to adapt and learn. But we are still left with a big mystery here. What is stopping B cells from generating antibodies that attack our own self-cells? How do they know the difference between an invading pathogen's antigen and a self-antigen? The protein molecules that our cells produce and the ones produced by pathogens are almost identical, after all.

Perhaps not surprisingly, some B cells *do* create antibodies that attack our self cells. And because their antibodies stick to self-antigens, you might expect them to clone themselves and spread—unrest leading to civil war within our bodies. But thankfully this is rare. Instead, it turns out that immature B cells commit suicide before allowing that to happen.

When new B cells are freshly made in the bone marrow or while waiting in the lymph nodes for a signal from the helper T cells, they are surrounded by our self cells and self-antigens. If any of these self-antigens sticks to an immature B cell, there will be a danger of the cell churning out harmful antibodies when it matures. Instead, it dies. B cells are pre-programmed to commit suicide if they encounter antigen while still immature. And B cells mature only when they receive the right chemical signals. What provides the signals? Those clever helper T cells do.

Although the details of this process are fiendishly complex (with many of the finer points still to be worked out by scientists), it seems that "cell suicide pacts" provide one important way for our immune system to maintain self-tolerance. Those parts of it that may attack self will die, be killed, or be suppressed by various other parts. This process of removing cells that attack us is known as *negative selection*.

In addition to the evolution, memory, and negative selection of B cells, much the same kinds of processes go on in our thymus (a small organ in our chests) for T cells. These and many other processes ensure that our macrophages, B cells, and T cells all work together, attacking invading pathogens but not self-cells.

WHEN THINGS GO WRONG

Your immune system is the reason you are alive and able to read these words. It protects you and cures you of almost every virus, bacterium, parasite, and fungus that you inhale, eat, or incur through injury. We know this, for we have seen the tragic consequences when people do not have functioning immune systems.

In a condition known as severe combined immunodeficiency disease (SCID)—made famous by David, the "bubble boy"—patients are born with defective genes that prevent the correct development of B and T cells. The only hope for survival used to be complete isolation from the outside world to prevent any form of infection. With no immune system, even a minor infection would kill someone with SCID, so all pathogens have to be kept from entering their bodies. David's story is tragic: he lived to the age of twelve inside a large tentlike bubble with special air filters. Eventually, a bone marrow transplant from his sister was attempted. David died in February 1984, 124 days after the operation, from a normally harmless virus unknowingly transferred within her marrow.*

The prospects for sufferers of SCID are improving. One of the most successful treatments has been gene therapy, where the DNA of a virus is modified to contain the missing genes of the patient. The virus is then used to infect the immune cells deliberately, where it inserts the missing gene into the DNA of the cells, causing them to function normally. By

---

*The virus was the Epstein-Barr virus—normally capable of causing nothing more serious than mononucleosis, or the "kissing disease." In patients with immunodeficiencies, the virus can infect B cells, causing uncontrollable proliferation (cloning), resulting in lymphoma—cancer of the lymph nodes.

correcting the DNA of stem cells within the bone marrow of patients (which makes all of the blood cells), we may be able to cure such SCID sufferers for life. But gene therapy research is still ongoing. Today, the most common treatment is still a bone marrow transplant.

Lacking an immune system because of a genetic disorder (known as a primary immune deficiency) is thankfully rare. Secondary immune deficiencies, or acquired immune deficiencies, are considerably more common. These conditions arise after exposure to drugs or chemicals (used to remove cancers or to prevent organ transplant rejections) or as a result of a disease. One of the most recent and most devastating of such diseases has been AIDS.

AIDS

AIDS is not truly a single disease; it is a syndrome—a collection of diseases typically caught as a result of contracting HIV (the human immunodeficiency virus). HIV, like all viruses, can infect only certain types of cell in our bodies. Unfortunately, HIV attacks the very cells designed to stop viruses: it infects T cells. Even worse, this virus is a retrovirus—it can convert its RNA into DNA and insert it into the DNA of the cell. Once there, the cell "thinks" that the instructions for producing HIV are just another part of its normal program, so will construct new copies of the HIV invader as part of its normal activities. The virus is also insidious in its timing: it becomes dormant for days or months before finally causing the infected cell to follow its instructions. It also produces a very prolonged and slow deterioration of T helper cells, which may take three to eight years, during which the patient shows no external symptoms. Eventually, lymphadenopathy (swollen and tender lymph nodes) occurs, followed by the onset of a number of opportunistic infections. These take advantage of the now-devastated immune system. Patients soon develop serious fungal and viral infections, leading to wasting, cancers, severe neurological disorders, and finally death.

As horrific as this sounds, the treatments for this disease are now impressive, and it seems that being HIV positive may no longer mean certain death by AIDS. The use of combination therapy, which combines many different drugs to inhibit or remove HIV from the cells, allows the helper T cells (and hence the rest of the immune system) to recover and remain stable. New drugs that prevent HIV from inserting its genes into the DNA of T cells have also shown great promise. While they are administered, 99.9 percent of the virus can be removed. However, there is still no cure. Patients must currently take these drugs for the rest of their lives if they wish to prevent the onset of AIDS.*

None of these conditions can be blamed on our immune systems. As always, this network of cells, organs, and chemicals within us tries to do its best to keep us alive, sometimes against insurmountable odds. But occasionally our protector can turn against us. Through overexcitement or stupidity, our immune systems sometimes kill us.

CONSUMPTION

Tuberculosis (TB) is no longer the familiar and hated enemy that killed one in ten Englishmen in the early nineteenth century.† Consumption, as it was then called, is a nasty disease that affected the lungs of patients and lingered for years, sometimes decades, before finally striking down its victims. Indeed, my great-grandfather suffered from TB in his early twenties, just around the time he was engaged to be married. He was told to move from his home in Wiltshire, England, to somewhere with a hot climate. He and his fiancée moved to South Africa, where thankfully he recovered. If he had not, you would not be reading this book, for my grandmother would not have been born, nor my mother, nor me.

But not all sufferers of consumption were so lucky. It was quickly discovered that TB was caused by particularly nasty mycobacterium called tubercle. These are airborne and enter our bodies via the lungs. We have macrophages that cluster at the lungs for exactly this eventuality, so the bacteria are immediately eaten by our "big eaters." This is often enough to destroy most bacteria, but tubercle are sometimes able to live inside macrophages like parasites, reproducing while hidden from sight.

---

*At the beginning of the year 2000, the Joint United Nations Programme on HIV/AIDS reported that "33.6 million people are estimated to be living with HIV/AIDS. Of these, 32.4 million are adults. 14.8 million are women, and 1.2 million are children under 15."

†But it has reappeared in more recent years, particularly in those with AIDS.

As we know, killer T cells are able to cope with such sneaky threats, so the typical response would be for our killer T cells to kill the infected macrophages. Unfortunately, this only has the effect of allowing the bacteria to escape and infect the surrounding lung tissue.

And it is at this point that our own immune system can turn against us. In an overenthusiastic attempt to remove all cells infected by the bacteria, our T cells begin killing large numbers of lung cells. In the final stages of this disease, much of the lung tissue may have been dissolved—not by the bacteria but by our own immune system. Consumption is an apt name, indeed.

Such self-inflicted, or autoimmune, diseases are distressingly common. Viral hepatitis, a disease that affects more than 300 million people worldwide, is another example. The liver becomes infected with the hepatitis B virus (HBV), and in the most chronic form, our immune system can destroy most of the liver while attempting to clear the infection. The result is cirrhosis, often leading to liver failure, cancerous growths, and death. To make matters even worse, it seems that HBV does not even affect the working of our liver cells. Our immune systems can kill us even when there is no threat to our health from a pathogen.*

NEW IDEAS OF IMMUNITY

It can be distressing to hear of such illnesses, but they teach us much. In the same way that brain disorders have helped us understand the working of brains, immune disorders and diseases have helped shape our understanding of the immune system. Today we know a great deal about the cells of the immune system and the chemical messages they use to activate and suppress different parts of each other. New ideas about how so many different elements are coordinated are now emerging.

THINKING ABOUT HEALTH

One startling finding has been the discovery of communication between the immune system and the brain. It has long been known that our immune systems can use chemicals to tell the brain to raise the body

---

*William Clark gives many more examples of the faculties and failings of immune systems in his excellent book *At War Within*.

temperature or alter blood flow by dilating blood vessels. More recent research has also shown that the brain is capable of remarkable control over the immune system. When patients are depressed, anxious, or stressed, their immune systems will also be depressed, rendering them more susceptible to illness. For example, recent widows and widowers are more likely than other people to fall ill, and their illnesses are more likely to be serious. Identical twins, both with a genetic tendency toward the autoimmune disease rheumatoid arthritis, can show extreme differences, corresponding to their personalities. The relaxed, easygoing twin may not suffer at all, while the nervous, high-strung twin suffers crippling arthritis.

Perhaps the most remarkable research in this area was performed with rats by professor of psychosocial medicine Robert Ader in the 1970s. Ader administered a powerful immune-suppressant drug in saccharine-flavored water to his rats. Within hours, their immune systems were unable to produce antibodies for foreign antigens. They also experienced side effects of nausea and vomiting from the drug, which were so unpleasant that the rats soon learned to associate the saccharine flavor with these side effects. Like Pavlov's dog, which salivated at the sound of a bell, the rats became nauseous just from the taste of saccharine-flavored water, even when the drug was not there. But the most remarkable finding was that the immune systems of the rats also became suppressed, in exactly the way the drug had been suppressing their immune systems. Their brains had become conditioned to reproduce both the side effects and the actual effects of the drug, even when it was not administered.

Since these findings, the same results have been obtained for humans. The experiments demonstrate just how much unconscious control our brains have over the functioning of our immune systems. Today we believe that B and T cells can respond to neurotransmitters and neurohormones produced by the brain. The brain can also respond to interleukins produced by the cells of the immune system. In addition, it seems that the brain can actually manufacture many of the chemicals used by the immune system cells for communication, and the immune system can make many neurochemicals. As William Clark says, "It turns out that the mind and the immune system are completely fluent both in their own language and in the language of the other. . . . It is almost as if the immune system were a chemical extension of the brain floating around in the bloodstream."

RESPONDING TO DANGER

But while we are making leaps and bounds in our understanding, there are still large gaps in our knowledge of this "chemical brain" within us. Earlier, we explored how negative selection using "cell suicide pacts" gave us self-tolerance. If you recall, our immune cells do not attack self-cells; they commit suicide before that can happen. Unfortunately, we do not fully understand how these mechanisms work. Why is it that our immune systems do not attack all of the foreign proteins we eat? Why does the immune system of a mother not reject the tissue of a growing fetus? Why are the helpful bacteria within our gut not attacked? Difficult questions such as these have led some researchers to create a new theory.

The somewhat eccentric Polly Matzinger and others developed an idea known as the *danger model* to explain self-tolerance in a different way.* Instead of relying on cells to know the difference between self and nonself (committing suicide if they might attack self), Matzinger suggests that our immune cells respond to danger signals. It turns out that there is a difference between programmed cell death ("suicide," or *apoptosis*) and abnormal cell death caused by infection. If cells are killed by bacteria or parasites, Matzinger suggests that they release a cry of pain in the form of a danger signal. Through chemical communication, this signal triggers the immune cells to act, causing the immune response.

Although the danger model is still controversial (having had a very mixed response from other scientists), current research shows that it may be at least partially true. This idea is unlikely to replace the conventional ideas of negative selection, but it does seem to be complementing the textbook view. As is normal in science, novel ideas are often debated by extremists on each side, and the truth is somewhere in the middle.

---

*Known for her strong personality and love of line dancing, Polly has an unusual background. Before discovering science, she had a series of jobs: jazz musician, *Playboy* bunny, dog trainer, brick cleaner. Eventually, she was working in a bar when she met a couple of scientists. Through their influence, she enrolled in a university and eventually became a scientist herself.

Immunologists are beginning to recognize a danger signal response as another ability in the growing repertoire of our immune systems.

NETWORKS AND AUTOPOETRY

In her writing on the danger model, Polly Matzinger describes the immune system as "not a separate army protecting (and regulating) the rest of the organs of the body, but an extended, highly interactive network making its decisions on the basis of input from all bodily tissues." Such views are not new in this field.

Since the early 1970s, the idea of an immune network has lurked in the background of immunology. The concept revolves around ideas of suppression by the immune system and was first proposed by Nobel Prize–winning immunologist Niels Kaj Jerne, who also helped to discover clonal selection some twenty years previously. Basing his theory on experimental results, Jerne proposed that immune cells can attack other immune cells. So if some B cells began manufacturing antibodies harmful to self-cells, other immune cells would attack and kill them, suppressing the autoimmune response. This can be thought of as a conceptual network, with free-floating cells connected to each other in the network if they can affect each other (via chemical links rather than static physical links). Unfortunately, perhaps because this network theory considered only antigen receptors (and receptors to receptors), it was quickly discredited when many more complex mechanisms were discovered in the immune system.

Nevertheless, ideas of other, larger networks within the immune system have not died. Inspired by Jerne's network theory, people such as cognitive neuroscientist Francisco Varela and physicist/systems theorist Fritjof Capra have developed similar views. Their approach is considerably more philosophical. Philosophers have great fun trying to explain consciousness and the brain; they also enjoy themselves with the equally mystifying abilities of the immune system.

Rather than network theory, Varela describes the immune system as an *autopoetic system*. The ideas do not suggest new mechanisms or new abilities of the immune system, they offer a different perspective on the way it works. According to this view, the immune system is no longer a collection of cells and organs that respond to invading pathogens to protect the body. Instead, it is a closed system of complex interacting elements that tries to maintain its delicate internal balance. When a pathogen invades, the balance is upset—and like a tightrope walker given a push, the immune system wobbles around until it regains its stability once again. Those wobbles are what we see as immune responses to pathogens, but really the immune system is trying to maintain its equilibrium, not defend against infection. The wobbles that it uses to recover its balance usually work, because the immune system is said to be *structurally coupled* to the environment and possible pathogens. Like two glass lenses ground together until they fit perfectly, all bumps and grooves meshing, evolution has "ground" the immune system and pathogens together until they fit perfectly.

Whether these ideas of autopoesis will turn out to be helpful or not, it is now becoming increasingly clear that the immune system is similar to the brain in many ways. Both have complex networks of interacting cells. Both use signaling to activate and suppress the action of their elements. And both display higher-level abilities such as learning and memory.

Perhaps, in a strange kind of way, your immune system has its own thoughts. I wonder, if it could sleep, would it dream?

COMPUTERS CATCH COLDS TOO

In late May 2000, an epidemic swept the world. Like many thousands of other people, I had firsthand experience of the disease, but luckily I was not infected myself. No biological organism was. Only computers were attacked by this menace.

The culprit was an innocuous-looking e-mail with the subject heading "I love you." Usually sent from colleagues or friends, the message came with an attached file. When the e-mail was opened and read, the file, which was really a program, sprang into action. It first wormed its way into the heart of the computer's operating system, ensuring that it would be executed every time the computer started. It then read the passwords used to gain access to the Internet and e-mailed them to an address in the Philippines. It trawled through the computer, moving, renaming, or overwriting multimedia files. If the computer was networked to any others, this program made copies of itself on all of the computers it could reach. But its final trick was the cleverest. It searched out the e-mail address book of the user of the computer. The program then masqueraded as the computer's owner and e-mailed copies of itself to everyone in the address book, with the subject "I love you." When the recipients of the message saw the compelling subject line and also saw that the e-mail had been sent from someone they knew, they would open and read the e-mail. And this set the program free on yet another computer.

This is an example of a computer virus—in this case a special kind of virus known as a worm. The "I love you" virus caused quite a stir for a few days. Many companies temporarily shut down their e-mail systems, and some lost important files. The cost of file restoration and wasted time was substantial, worldwide.*

Typically written by unscrupulous young programmers, computer viruses are nothing more than special kinds of programs. But although they exist solely within the digital universes of our computers, they share some profound similarities with biological viruses. Both types of virus hide within other entities, unable to become active without the help of their hosts. Both do their utmost to avoid detection. And both rely on deception to reproduce and spread.

For example, a file infector virus copies itself to the end of a program file, ensuring that its instructions are followed in addition to the instructions of the program. A boot sector virus secretes itself onto the special parts of discs that computers always read first, when starting. A Trojan horse pretends to be a useful program, but secretly performs some malicious activity (such as leaking passwords) when run. A macro virus hides within text documents, becoming active by making use of the macro language of the word processor when the text is read. Polymorphic computer viruses can even modify parts of themselves to try to avoid detection.

Like the biological viruses, the main goal of computer viruses is to duplicate and propagate to as many new hosts as possible. Also like biological viruses, computer viruses can cause considerable damage and upset to their hosts.

---

*By some considerable irony, while writing this I suffered a similar loss, so this is the second time this section has been written. The first version was completely obliterated, although the incident was due to poor software rather than computer infection—I think!

ANTIVIRUS IMMUNITY

Computer viruses are so prevalent that most new computers are sold with antivirus software as standard. This software forms the immune systems of our computers, designed to prevent, detect, and remove all virus infections. And much of the software unknowingly duplicates many of the tricks used by our immune systems.

Like the different jobs performed by the various immune cells, there are various types of antivirus software that tackle computer viruses in different ways. There are three main categories, according to the Computer Virus Industry Association. Infection prevention products (Class I) prevent infection or halt the replication of viruses. They often achieve this by preventing certain types of activity on the computer, such as writing to the boot sector or modifying program files. This is rather like the cocktail of chemicals and macrophages that fill our bodies (our *innate* immune systems), preventing simple pathogens from getting a foothold. Infection detection products (Class II) keep an eye on the files within the computer and notify the user if something suspicious has occurred (a program file changing its size, for example). This is very similar to the job of helper T cells as they examine the surfaces of macrophages to find out if infection has occurred. Infection identification products (Class III) check all the files within a computer for specific viruses. They usually search for unique identification strings that are characteristic of different viruses and either remove the virus from infected files or remove the infected files entirely. Again, this is surprisingly similar to the action of T and B cells, which identify pathogens using their unique antigen bar codes, then remove them using tailored antibodies or destruction of the infected cell.

DIGITAL PARASITES

So computers suffer from viruses just as we do. But this is not the only kind of illness that can befall them. There are more malevolent diseases that can be contracted by unwitting computers. Known by computer security officers as intrusions, the most common name for these parasites is *hacker*.

The word *hacker* used to mean someone very proficient at using a computer. Today, its most common usage refers to someone trying to break into a computer system. Because computer systems are wired together in networks and those networks are linked by telephone lines to form the Internet, it is possible for programs on one computer to access data on other computers anywhere in the world. Anyone with the right knowledge and ability can gain access to data stored on other computers. Hackers can read other people's e-mail, discover salary information, modify financial transactions, alter exam results, delete criminal records. Illegal intrusions into computer systems can cause extreme damage and financial loss to companies and individuals.

Companies spend millions each year in an attempt to foil such electronic thieves. Physical security is the first defense, with closed-circuit cameras, locked doors, security passes, and isolated computers. Like our own skin, the best protection against intrusions is to stop them from getting to the system they're trying to invade. But if they find a way in (often through the telephone line), what then?

The second line of defense is usually passwords. Unless you know the right password, you can't gain access; you need the right key for the digital lock. This is a bit like the chemical coating that a biological virus needs in order to gain entry into one of our cells. You can get in only with the right code—or at least that used to be true until World Wide Web pages were invented. Web browsers made it possible for people to access private files via the Internet server (the computer that serves the Web pages). To overcome this loophole, software called firewalls had to be invented. These programs place software barriers against intruders from outside, allowing access only from authorized users.

And yet the hackers still gain entry. The trouble is, like our bodies, computer systems are immensely complicated. A computer network will have vast amounts of different software running on it, many connections to the Internet, and often links to other local networks. All of this complexity provides new routes in for the hacker. Known bugs in mail programs may be used, hackers may masquerade as legitimate users, and any number of other tricks may be employed. (I'm being deliberately vague here.)

INFORMATIC SECURITY

Rather predictably, the problems of illegal intrusion have necessitated the creation of new software: intrusion detection systems (IDS). Like macrophages clustering around a wound, an IDS sits on the network connections, monitoring the flow of data through the computer network. It typically maintains an audit of activity: network data flow, log-in details, which programs or processes are being executed where. Once it has a pretty good idea of what is going on, the IDS tries to spot whether any of the activity is being caused by an intruder.

It usually employs one of two methods: try to spot a familiar hacker pattern or try to spot something abnormal. The first method (known as a misuse detector) works rather like the Class III virus detectors: it has a list of different hacker patterns, so if it finds one in its audit data, it's found a hacker. The second type of IDS (known as an anomaly detector) is a little cleverer. This software builds up a picture of what normal, everyday activity on the computer system looks like. If anything happens that deviates from this norm, it alerts the person in charge. The idea is not unlike Polly Matzinger's danger model of the immune system: rather than look for specific threats, respond when normal behavior is disrupted.

The most advanced intrusion detection systems combine misuse and anomaly detection into one. But even then, there are risks from hackers. Like the rapid evolution of biological pathogens, hackers and viruses are constantly updating their methods to overcome detection. And so the ongoing problems have led researchers to investigate the immune system. Perhaps processes within our own immune systems could be used to stop infection of and intrusion into our computers . . .

HOW TO MAKE A GOOD DIGITAL IMMUNE SYSTEM

Computer viruses and biological viruses. Hackers and parasites. Antivirus software, intrusion detection software, and immune systems. The similarities are surprising. But despite the parallels, there are some significant differences between traditional virus and intrusion detection systems and our immune systems.

Perhaps the most important difference is location. Our immune systems are distributed throughout our bodies. Even if we lose entire limbs or organs, the immune system will still operate. But antivirus and intrusion detection software is normally in a single place, on one computer. Should that program crash (or be removed), then the computer (and often an entire network of computers) will become vulnerable.

There are other problems, too. Much of the antivirus and intrusion detection software relies on having prior knowledge of enemies. The software looks for patterns (virus or hacker), but if the digital pathogen is new or different, the patterns will not match, and the infection spreads unchecked. But our immune systems are not like this. They learn the new antigen patterns and learn how to respond.

IBM'S PETRI DISH

The idea of learning within immune systems was not lost on researchers at IBM's Watson Research Center in New York State. Inspired by the ability of immune cells to identify and learn the antigens of new pathogens, they built a computer system capable of doing something analogous for computer viruses.

The program uses a combination of anomaly and misuse detection for viruses to try to spot whether anything unusual is happening. If it discovers something suspicious, it takes the potential virus and places it on a "digital petri dish." In controlled conditions, the potential virus is exposed to some decoy programs—if the programs become infected, then IBM's antivirus software detects the change. The program then compares the difference between the uninfected programs and the infected ones and derives the virus pattern. This is added to the antivirus software's store of knowledge, allowing the new virus to be detected in the wild.

It's a clever idea, making use of computer learning to enable antivirus software to update itself, without the need for human experts to extract viral patterns. Rather like the action of clonal selection and memory cells, IBM's software can discover and remember new patterns of infection. But although the software was doing similar things as our immune systems, the IBM researchers actually used a different form of digital biology to produce these abilities: neural networks. Other researchers have taken the idea of computer immune systems a few steps further.

COMPUTER IMMUNE SYSTEMS

Immune systems are complicated—in many ways, even more complicated than brains. We can usually see what brains do (make us walk and talk, for example), but it has often been very hard to work out what immune systems do. We simply don't get to see the activities of our immune system on a daily basis. And when we delve into the biology, we discover terrifying terminology and confusing cells that all do different jobs in different ways—all of which means that very few computer scientists have had the courage or persistence to develop digital immune systems within computers. Unlike the fields of neural and evolutionary computation, the use of processes from immune systems has just started. This is a field in its infancy, barely ten years old.

Nevertheless, computer immune systems are now sprouting within computers in research laboratories around the world, and the most influential of the groups in this area must be the computer immune systems team at the University of New Mexico, Albuquerque. Led by Stephanie Forrest, the researchers here have pioneered the ideas of using immune processes within computers. They are a multidisciplinary group, working with immunologists such as Alan Perelson to develop computer algorithms that are faithful to our understanding of how immunity develops inside us. Stephanie explained how this work began: "I was introduced to the wonders of the immune system when I was carpooling to Los Alamos National Laboratory with Alan Perelson, a theoretical immunologist. Initially, I was intrigued by the fact that the immune system is highly decentralized and massively parallel, and I wanted to understand how it could achieve such sophisticated pattern recognition in such a distributed way."

Stephanie knew that a number of different processes occur to create abilities such as diversity, specificity, memory, and tolerance. They allow immune systems to be distributed throughout our bodies, to learn and adapt. She and the other researchers in the group carefully isolated these different processes, transplanting them into the digital universes of our computers with the skill of surgeons. They use the same processes to manipulate digital antibodies spread throughout computer networks. The computer immune systems continue to work even if one piece of software or one specific computer is turned off.

One of the first processes they transferred into computers was generation of diversity.

DIGITAL GODS

Diversity of antibodies is essential to make sure that our immune systems can always find an answer to the strangest of new antigens. Diversity of digital antibodies can be equally important, particularly if the digital antibodies are computer hacker detectors.

One strand of research has focused on the GOD (generation of diversity) question for digital antibodies. In biology, antibodies are constructed using the candy store "pick and mix" technique—different fragments are combined to produce unique combinations. Over millions of years, our DNA has evolved to ensure that the right fragments are available for this construction job. Certain collections of genes (known as gene libraries) specify the protein fragments that can be used to assemble new antibodies.

But this presented a tricky question to computer scientists. Could digital gene libraries be evolved that would permit sufficient diversity for digital antibodies? If they could not, then there would be no easy way to create the huge numbers of different digital antibodies one would need to detect computer hackers or viruses. We cannot simply generate random antibodies, for then far too many will mistakenly detect self (normal programs that are allowed to run on the computer). We need the knowledge contained within the gene libraries to help create diverse and useful digital antibodies.

After some research, Stephanie and her team discovered that the answer was "yes"—evolution within our computers can produce excellent gene libraries, which can then be used to generate a diverse and useful range of digital antibodies. They achieved this result by using a genetic algorithm—something that should be familiar from the chapter on evolution. As Stephanie puts it, their work used "an abstract universe based on binary strings in which aspects of the immune system are modeled by interactions among the strings." In other words, they created a digital universe in which strings of binary 1s and 0s were used to form gene libraries, antibodies, and antigens. Digital individuals in this universe had immune systems that made large numbers of antibodies using their own gene libraries. The more successful the individual's antibodies were at detecting digital antigens (hackers), the fitter the individual was. And the genetic algorithm evolved the individuals based on their fitness. The result was that evolution adjusted the gene libraries of the individuals to ensure that each could produce extremely diverse antibodies. The genetic algorithm evolved good immune systems within the computer.

Other researchers have also investigated digital antibody diversity. Through some imaginative tangential thinking, they have suggested that antibodies can be thought of as solutions to problems. The immune system creates a diverse range of possible solutions, which can then be fine-tuned into a single, good solution. So if you change the problem from detecting hackers to something else, the immune system can solve the new problem. As bizarre as it sounds, like the ideas of swarm intelligence in a previous chapter, it works. Researchers have used principles of diversity generation for function optimization, showing that immune systems can solve pure mathematical problems as expertly as they keep us healthy.

EVOLVING PRECISE DIGITAL CLONES

So we can evolve digital diversity. Could anything more from our immune systems be used within our computers? Stephanie Forrest thought so and created another approach, called the *clonal selection algorithm*. We saw how B cells that generate better antibodies for a given antigen will clone themselves—like the music industry when it clones bands. With an occasional mutation to vary the antibody and an occasional development of a memory cell, our immune systems make their antibody response specific and long lasting. They're using evolution to adjust themselves inside us.

The clonal selection algorithm works in exactly the same way. Digital antibodies (or solutions) are generated and evaluated. The better ones (that match digital antigens or solve a problem) clone themselves a little more often than the worse ones. With an occasional mutation to vary the antibody, the algorithm is able to evolve good antibodies (or solutions).

If you're very perceptive, you may have noticed that this sounds exactly like a genetic algorithm. And indeed it is—except for one important difference. The evaluation of antibodies happens in a special way. Imagine you're an antigen floating around. You may come into close contact with a number of antibodies, but you will stick to the one that is most specific to you. On the other hand, imagine you're an antibody. You will be specific to only a few antigens, but you may not ever get to see that antigen. You may only see other antigens.

The clonal selection algorithm duplicates these uncertainties. When evaluating digital antibodies, it picks a random group of them and exposes them to a random antigen. The antigen sticks to the best matching antibody, so the fitness of that antibody alone is increased. And then the process is repeated with a different random group of antibodies and another random antigen. And then repeated again and again.

The result of this way of evaluating antibodies is that niches develop as evolution progresses. Separate groups of specialists form, each containing experts at detecting different antigens. In computer science, this made researchers sit up and take notice, for niching in evolutionary algorithms had always required complicated additions called fitness sharing. Somehow the clonal selection algorithm was able to create and maintain niches through natural evolution alone, using a simple and elegant procedure.

Since its creation, this algorithm has been shown to work very successfully as an evolutionary algorithm in its own right. It can detect hackers and viruses, and like the other evolutionary approaches, it can solve a million other problems too.

DIGITAL TOLERANCE BY DELETION

Diversity and clonal selection are all very well, but our immune systems do so much more. One rather significant thing that they do is recognize the difference between self and nonself, attacking only foreign invaders. One important way that our immune systems achieve this feat is by employing cell suicide pacts (negative selection): cells die rather than attack self.

Again, Stephanie Forrest and her team realized the significance of negative selection in biology and decided to try it within their computers. They developed the negative selection algorithm to check out the potential of this process. The algorithm is deceptively simple: keep generating digital antibodies, and if any match self, delete them. That's it.

The program turned out to be pretty good at general machine-learning problems in computer science, particularly because it works in a back-to-front way. Normally, if we are using the computer to learn how to find something hidden among lots of data (like the trace of a virus among all the files in the computer), we say, "Keep detectors that are good at detecting viruses and throw away everything else." But the negative selection algorithm says, "Throw away detectors that match normal things, and keep everything else." So after running these programs for a while, you end up with a set of detectors (digital antibodies) that detect everything that is not normal. You have an anomaly detector—a program that will discover abnormal things such as viruses and hackers—even if they are brand new and have never been seen before. Negative selection algorithms have been demonstrated for intrusion detection on simple computer networks, successfully finding simulated hackers. When this algorithm is combined with the other methods (such as gene library evolution), it will hopefully become usable for live real-world intrusion detection.

But not all types of digital immune systems are so disparate and esoteric. Let me introduce you to a pet of ours at University College London, where I work.

NEW IDEAS FOR DIGITAL IMMUNITY

The lungs of the creature create a constant flow of air, in and out. The incoming air stimulates the organs of the creature, helping them to do their various tasks. As air leaves the creature, it carries with it the waste products from various organs. But although this life-bringing substance for the organism is essential to allow it to function, the air sometimes contains malevolent entities. These intruders attempt to infect one or more organs of the creature, preventing them from carrying out their proper tasks or even killing them.

Thankfully, like all complex organisms, this creature has an immune system. New cells are constantly created and checked in its bone marrow and thymus to counter the threat of disease. Each produces a special antibody detector, designed with the help of gene libraries within the creature to ensure sufficient diversity. Not only that, the gene libraries hold information about which kinds of antibodies are usually most useful to make, ensuring that most new antibodies are effective. However, if any of the antibodies matches any self-antigens of the creature, they are destroyed, leaving a collection of antibodies that detect nonself. These flow through the organs of the creature, maintaining a vigilant watch for intruders.

When an antibody detects an antigen, it sends out a signal: "I've detected something abnormal." Other cells of the immune system then investigate the entity that has been detected. If it turns out that this was indeed a potential threat to the health of the creature, the numbers of the antibody are increased as new clones are made. These antibodies are then fine-tuned, for some clones have slight mutations that result in variations to the original antibody. As the worse antibodies detect fewer of the intruders and die of old age, the remaining antibodies become better and better at detecting the current type of intruder. Even after the current intruder has been removed, the immune system does not forget its lesson. By creating long-lived antibodies that persist in the organs of the creature, it responds even faster and more accurately next time.

As you may have guessed, the creature exists inside a digital universe, this time constructed using a network of computers. The "air" that the creature "breathed" in and out is the flow of data known as network traffic from other networks (such as the Internet) to the computer network and back. The organs of the creature are different pieces of software spread across separate computers on the network. So software modules that perform the job of gene libraries, lymph nodes, thymus, and bone marrow all exist on various computers. The creature's antibodies are digital detectors that migrate between computers on the network, searching for antigens (intruders). And the antigens are summaries of network traffic. So every data item (or packet) transmitted to or from the network becomes represented as an antigen, and the digital immune system keeps its distributed eyes out for any antigens that look abnormal. If it finds one, it has found someone hacking into the network or misusing the computers in some way.

This creature has not been born yet. It is still under development by my doctoral student Jungwon Kim.* Jungwon is investigating how processes of our immune systems can be transferred to real-world intrusion detection and fraud problems. She is one of the first researchers to investigate the combination of diversity generation, negative selection, and clonal selection in a single computer immune system. Our hope is that the combination of all three processes will lead to computer immune systems with more of the abilities of our own immune systems. But Jungwon is motivated by very practical concerns. Like many other computer scientists, she is more interested in solving problems than in biology. As she said, "Our traditional algorithms were originally devised to solve rather small problems. When we try to solve real-world problems, the algorithms are modified so much that they can become unwieldy. On the other hand, evolution has already designed biological mechanisms to work for complicated systems. So using ideas from biology for solving our complicated real problems might be a better starting point."

Although there is still much work to do, this kind of research is now showing us just how effective the processes of diversity generation, negative selection, and clonal selection can be for intrusion detection in computer networks.

FROM DIGITAL TO BIOLOGY

Other researchers now examine how immune systems can solve many other problems, such as scheduling, pattern recognition, aircraft control, design of truss structures, traveling salesman problems, data mining, noise reduction, diagnosis of sensor readings, and even satellite image recognition. Combinations of immune systems and neural networks (immunized neural networks) provide improved learning for many neural network problems. Meanwhile, Stephanie Forrest and her team are becoming more involved in the use of computers to understand biological immune systems. Because the actions of immune cells are so complex, it seems that our only hope of truly understanding them is to create digital versions and watch what they do.

---

*Who, I should point out, does not get quite so carried away with natural analogies and metaphors as her supervisor . . .

She is not the only one. Robin Callard at the Institute of Child Health in London is an immunobiologist with similar views. When we last spoke, he told me why he felt this way: "I started to read some popular science books on chaos and complexity ... I suddenly realized that biological systems were actually quite different than most biologists appreciated. Most biologists think of biological systems as being a jigsaw. You've got this view where each bit of information slots into place, and then all the bits of information together help us to understand how it works. But only a casual reading of someone like Kauffman shows that this is not the case. So I thought it was time to do something about it."

Robin was frustrated at the lack of progress in immunobiology. Although researchers were identifying the actions of individual cells and proteins, no one completely understood how everything worked together. So he decided to join the increasing numbers of biologists who start up collaborative projects with mathematicians and computer scientists. He and his group now model specific behaviors of our immune systems using nonlinear equations on a computer. Other groups use cellular automata (which we explore in the next chapter).

DIGITAL NETWORKS AND AUTOPOETRY

Strangely, many computer science researchers have developed computer immune systems that are based on Jerne's network theory. Little robots are controlled by network immune systems. Genetic programming, intelligent buildings, multimodal optimization, and many other applications have been created according to Jerne's idea. And yet the network theory is an idea that "no card-carrying immunobiologist would subscribe to these days," according to Callard. It may seem strange to build digital biology that works according to discredited theories—but as such researchers say, "If it works, it works." Sometimes digital biology does not have to follow processes identical to nature to be useful.

Even the somewhat confusing theories of autopoesis are inspiring new creations of digital biology. Often going under the name of *embodiment,* ideas of self-maintaining networks of robots structurally coupled to their environments keep researchers entertained for many hours.

One such project is being attempted by Tom Quick, who has a strong background in philosophy and works with Kerstin Dautenhahn, well known for her ideas in this area. He is developing a digital creature that, like Varela's autopoetic view of the immune system, comprises a closed system of interacting elements that try to maintain their internal balance. When the balance is upset, the actions performed by the creature will be the result of its trying to return to its equilibrium state. By somehow structurally coupling the creature to its environment, these actions will perform useful activities. Tom hopes that the digital creature will be able to explore the Internet and that, when transplanted into the electronic brain of a little robot, it will enable the robot to move sensibly in its environment. In his own words, "I hope that because the dynamics of interaction between the structure of the system and its environment is similar . . . provided the relationships are similar, then you get similar behaviors emerging—just kind of translated across different domains in effect."

Like an increasing number of scientists, Tom is arguing that the same processes that are seen in our immune systems can be moved into other domains. Stephanie Forrest agrees: "I expect that when we understand how the immune system responds to foreign pathogens and how it regulates itself, we will discover new principles and algorithms that can be applied to computer science. Immunology still has much to discover about how natural immune systems work, and I would be surprised if none of what is discovered in the future could be applied to computer science."

From biology to robots to digital universes, if the idea is a good one, it will usually work wherever it is used. Like our brains, immune systems will continue to progress our technology for many years to come.

SUMMARY

Immune systems are the astonishing products of millions of years of organisms fighting to stay alive. Our immune systems, like our brains, are thought to exemplify nature's most advanced biological technology among life on Earth. Instead of the mysterious characteristic called consciousness, they maintain the equally hard-to-untangle notion of health. They keep us alive despite the relentless, never-ending attack from every imaginable microscopic fiend. And because of the sheer variety and diversity of our attackers, our immune systems have had to employ ever more complicated countermeasures.

The processes of our immune systems are impressive. Diversity of antibodies via "pick and mix" protein construction. Specificity of antibodies using the evolution of clonal selection. Memory by long-lived B cells. Self-tolerance through negative selection's cell suicide pacts. These mechanisms (and many hundreds of others that we have not explored) are the results of an eons-long arms race between us and our pathogens.

This advanced pathogen-fighting force within us has now become our teacher. We are learning how to fight computer viruses and hackers by employing the same processes that keep us alive. We are also learning advanced new ways of solving other problems. The chemical brains within us are proving to be very knowledgeable.

But if the immune system and the brain are clever, what about the processes that made them develop? Evolution may have generated the genetic instructions, but the processes of *growth* from a single cell to adult organism are not to be underestimated.

GROWTH

"THAT THE EXPRESSION of human genes must be a highly regulated process should be clear to anyone who has ever dissected a human body."*

You don't say.

It tickled my sense of humor when I read this initial sentence in a biology book recently. Clearly the author believed that many of his readers would have dissected human bodies. I must say that I rather hope that you, my reader, have not.

But, whomever his work was aimed at, I have to say that he is right about genes. The expression of our genes, and indeed the genes of every other organism on the planet, is an astonishing, cascading plethora of regulation. Our genes were written by evolution, nature's master programmer, and together form the most convoluted set of instructions we know of. More advanced than any computer program we have written, when the program of development runs, we are the result. Through the expression of our genes, the behavior of our cells is controlled. Cells change, communicate, move, multiply . . . and grow a whole creature, all controlled by genetic instructions. And it all begins with just one set of genes, held within a single cell.

That cell is the single fertilized egg that results from sexual reproduction. Whether you're a plant, an ant, or a human being, you began as a

---

*David Latchman, *Basic Molecular and Cell Biology* (2nd ed.). And my apologies to developmental biologists for the title of this chapter. I know it should be *Development*, but I like *Growth* better.

single cell, too small to see with the naked eye. To make you, the cell had to construct more copies of itself, those copies had to make copies, those copies had to make even more copies, etc. Vast sheets of developing cells were folded, pushed, squeezed, and even killed to mold everything into the right shape at the right time. While being placed in the right areas, the cells changed into liver, muscle, kidney, skin, immune, and brain cells. Cells were connected together into blood supplies, lymphatic systems, neural networks. Cartilage and bone were grown, blood cells created, limbs, head, eyes, and ears formed. Your heart was delicately started as the first signs of primitive brain activity began. As your muscles developed, you began to use them, kicking and moving in the womb. And the very process of using them helped them to develop further, ensuring their placement and growth was correct. After nine months (or thereabouts), one of the most complicated things in the universe—you—were born. Although still unable to move yourself about, you were equipped with the latest designs of brain and immune system, created to adapt and learn at a ferocious rate. And that's what you did. You learned faster than any mature human could; you grew stronger, faster, and more skilled; you encountered and survived attacks from countless viruses and bacteria.

Nine months to create such a mind-bogglingly complicated package of billions upon billions of cells. All from one, single cell. And all controlled by genes. It is a spectacular achievement of nature, which never fails to produce awe in parents. We often call it "the miracle of birth," but this is not entirely what we mean. A better phrase would be "the miracle of development."

This chapter explores this miracle that touches so many of our lives in such a moving way. We'll look at how growth (or, to use the correct word, *development*) creates us and every other multicellular organism around us. And our first step is to understand how our genes control and direct our development. The answer lies deep inside each of our cells.

ECLECTIC GENETICS

Over the previous chapters, we've seen plants and insects, brains and immune systems. They're all made from billions and billions of cells. And each of them grew from a single cell: the fertilized egg.

Like almost all of the cells in our bodies, the single new human cell formed by the fusion of sperm and ovum contains twenty-three pairs of chromosomes (half being contributed by each parent). These are all packed into a tiny sac inside the cell called the nucleus.

Our forty-six chromosomes are made from DNA molecules. These are constructed out of four building blocks, or bases: adenine (A), guanine (G), cystosine (C), and thymine (T). But rather than being tiny clumps of atoms like water molecules, DNA molecules are arranged in chains of bases (such as ACAAACT), forming long strings. These strings are so long, in fact, that if you were to stretch all forty-six chromosomes end to end, the total length would be 2 meters.

Two meters of DNA in each of our cells. I'm exactly six feet tall, and that's a half foot taller than I. How can it fit inside a nucleus so small that we cannot even see it? If we were to magnify the whole thing one thousand times, each chromosome would be a string 2 mm thick and 40 kilometers long. Altogether, there would be enough string to stretch from London to Naples, or from Manhattan to New Orleans. And we'd have to cram it into the space of a small bathroom. But we can't just jam in our string all higgledy-piggledy. We must be able to access different pieces of the string, for it contains important instructions. It tells us how to develop.

So how do we put our string into the bathroom? We *coil* it. As we twist the string, it becomes shorter, and soon it coils back on itself, collapsing into a smaller and smaller space. And this, of course, is what nature does. The enormous lengths of DNA fit inside the nucleus by a clever arrangement of coiling. Coils of coils of coils of coils of DNA compact the twenty-three pairs of chromosomes into microscopic structures. The structures are cleverly designed so that they can tuck away certain parts of the chromosome, exposing only those needed to guide the current cell to perform the current activity.

Although our chromosomes contain instructions to make us, not every part of our chromosomes has a function. A good two-thirds of all of the DNA is redundant, being composed of repeating patterns of bases that do nothing (that we are aware of). A good part of the rest is composed of more complex sequences that occur many times in different parts of our DNA. Only 5 percent of our DNA contains unique patterns (each often many thousands of bases long) that are actually used to control development. (It may even be as low as 3 percent.) These patterns are called *genes,* and we know they are used because we know that they define *proteins* within cells.

Only 5 percent of our DNA needed to define our development. And some scientists have worked out exactly how much information that is. A base pair can have any of four possible values: A, G, C, or T. So each base pair stores the equivalent of 2 binary digits of information. Because of some redundancy in genetic code, a base represents around 1.5 bits of information. This means that, in total, there may be only about 30 megabytes of information stored in our DNA that is actually used.

And that is unbelievably low. The word processor program I am using on my computer right now is much the same size. And although the word processor uses all of the complicated hardware and software that make up my computer—just as our genes use all of the machinery inside our cells in order to be expressed—my word processor is paltry and insignificant compared to the program of development. Evolution has discovered an immensely efficient and compact way of describing organisms. By using our genes to define how we develop, and not to define directly specific features (like length of arms, eye color, shape of ears), nature minimizes the amount of information that is needed. And the less information you need, the easier it is to copy it accurately into all of our cells.

So our genes are the instructions that describe how we are built. But how do they do it? They're only fragments of molecules, after all.

COMPLEX EXPRESSIONS

To understand, let's look a little closer at the first cell that was the start of your existence on this planet. The fertilized egg, or zygote, contained forty-six chromosomes, half from each parent. As the zygote traveled down the Fallopian tube toward your mother's womb, it began to divide, and divide again. Soon there was a cluster of cells, each containing an identical copy of your DNA. But the cells quickly began to look a little different. As your cells multiplied, they changed. Each produced slightly different cocktails of proteins, which caused it and its neighbors to organize its position, size, and function. Proteins changed your cells, and your cells made proteins.

If cells are musicians, then proteins are their music. Like the flow of jazz music, which dynamically affects the structure and organization of music played by each musician, the diffusion of proteins determines the organization and layout of cells in a body. Proteins change the behavior of cells just as music changes the behavior of musicians. And proteins are made from templates in our genes. As embryologist Lewis Wolpert said recently, "The way to think about development is it's cells changing their behavior. And that change in behavior is determined by genes. So genes determine how cells behave. They do this by determining what proteins there are in the cell."

In essence, that is what DNA is for: to make proteins. *Gene expression* is the process of transcribing the DNA in our genes through a conveyor belt of different chemicals, producing small chemical fragments that are assembled into proteins by cells. The proteins then direct our cells to change, move and grow.

So DNA is a set of protein assembly instructions for cells. The instructions are written in a language that has an alphabet of four letters: A, G, C, T. Genes form the words in the language.

But much to the frustration of scientists, these words are not written inside our chromosomes in the same orderly way that the words on this page are written. Unfortunately, genes are not written in any clear order, and because of all the nonfunctional DNA that gets in the way, each gene may be fragmented and interrupted by unused sequences of bases. Imagine if the instructions that came with your unassembled bookshelf looked like this: *xy 'yA' nyaxyilx pyiyece xyplayxce. gxyrooyxve Slxyide inxyto xanyd ixynto,* when the instruction was actually: *Slide piece A into groove and nail into place.*

When it's hard to understand which letters make up the words, the order in which the words should be read, and what they mean, it is almost impossible for us to understand the instructions.* But nature has

---

*It's one reason that the Human Genome Project is just the first step of a long journey toward understanding human development. We now know most of the sequence of DNA bases in human chromosomes, but we still have to work out where all of the genes are. And that's the easy part. Working out how all the genes coordinate their efforts to build a human being is still beyond our capabilities.

no such problem. The machinery within our cells is designed to untangle the DNA instructions within them.

As usual, it's a very clever process. The transcription of a gene begins when the right kind of chemical sticks (or binds) to the beginning of the gene on the DNA molecule. This unzips a short section, which becomes a template for the construction of a corresponding RNA molecule. During this process, any parts of the new RNA molecule that do not code for a protein (an intervening sequence, or intron) are spliced out, leaving an untangled messenger RNA molecule. We're not entirely clear how this is achieved, but it seems to be performed by smaller RNA molecules and proteins that form structures known as spliceosomes. They detect the end of one coding sequence (or exon) and beginning of the next and bind to them, forming a molecular loop that then falls away, snipping out the intron.* The newly transcribed RNA molecule then has a molecular cap and tail added to each end to make it last a little longer. Finally, the whole thing is transported from the nucleus of the cell into the cytoplasm, where it is translated into corresponding protein molecules. From DNA to RNA to protein, with untangling and added longevity, this is the process of gene expression.

REGULATING GENES

But that is far from the whole story. You may have noticed that earlier I said, "The transcription of a gene begins when the right kind of chemical sticks (or binds) to the beginning of the gene on the DNA molecule." In other words, genes are expressed only when a chemical triggers their expression.

There is a very good reason why this is required. Imagine if there were no triggers. If all genes were always transcribed all of the time, then every cell would pump out the full repertoire of proteins all of the time. And since every cell contains an identical copy of our DNA, they would all be the same. Produce the same proteins, and you become the same cell. The

---

*If this sounds like an overcomplicated way of arranging genes, there are benefits to be gained. By splitting genes among separate exons in our DNA, the same fragments of instructions can be shared, just as some of our words share common fragments: *them*, *their*, *other*. It's an example of a trick used by evolution to keep the size of coding DNA down.

only way cells can produce different proteins and become different from each other is if some genes are expressed in some cells and not in others.

So with the exception of certain housekeeping genes that produce proteins needed by all cells, our genes are conditional. In each new cell, some will be turned on, and some will be turned off. And like the code of a combination lock, each genetic combination of active and inactive genes unlocks a specific mixture of proteins, designed to transform the behavior and shape of the current cell. One genetic combination will make the cell into a skin cell, another will produce a neuron, another will make a B cell of the immune system. And what is the trigger that turns genes on and off? More protein molecules. These are transported from outside the cell into the nucleus, where they bind to specific regions of our coiled DNA, expressing or suppressing the genes.

If you're a computer scientist, like me, it all starts to feel spookily familiar, for our chromosomes are like computer programs. Although the instructions are fragmented and unordered, our genes use all of the programming tricks that we have created in our computer software. A gene may produce a protein that causes another to be expressed in a different cell. Or it may produce a protein that stops the expression of another. Clusters of genes can be expressed together. Genes may produce a protein for a specific amount of time, and stop. And if you know anything about programming computers, then you may have noticed that we have direct equivalents in our programming languages. IF . . . THEN statements, subroutines, loops, and even complicated techniques such as recursion are all used during gene expression

We can even write out a pretend set of genetic instructions in a computer language:

```
IF protein1 THEN express(gene1)
IF protein2 THEN express(gene3,gene5,gene6)
IF NOT protein2 THEN express(gene4)
express(gene1) { make protein2 }
express(gene3) { make protein5 }
express(gene4) { make protein6 }
express(gene5) { make protein1 }
express(gene6) { make protein7 }
```

In other words, if a cell begins life in the presence of protein 1, it will have its gene 1 turned on and will be a protein 2–producing cell. If a neighboring cell then encounters this protein 2, it will have three different genes turned on and will become a protein 5–, 1–, and 7–producing cell. And the protein 1 generated by this cell may diffuse its way back to the first cell, helping to keep it a consistent producer of protein 2. But if a cell some distance away does not have any protein 2 in its vicinity, then it will have its gene 4 expressed, and will become a protein 6 producer.

It does not take many genetic instructions to produce a complicated little program. If you had trouble following my example, imagine how complicated it is with tens of thousands of instructions, all causing cells to generate different proteins that activate or suppress the genes of other cells, which produce proteins that activate or suppress genes of yet more cells, and so on. Genes inside one cell regulate the genes of nearby cells, producing a massive and largely incomprehensible network of cell interactions.

Geneticist Steve Jones agrees. Describing his own work, he said, "My main concern with the public conception of genetics is that people think it is a much simpler subject than it is. And the assumption that there is simply a gene for this or for that is usually mistaken. . . . So my role, I think, is to try and demystify genetics and to make it *less clear* than it seems to be—because it's not very clear."

Of course he's right. There are not genes for long legs or even genes for blue eyes. There are only genes for proteins. From DNA to genes to proteins (via gene expression) to gene regulation (via protein interaction between cells). This is how our genes define us. As Steve says, "My feeling is that DNA may itself be quite simple, but once you step off the DNA you're in a morass of complexity."

And it's only the beginning of the journey from zygote to organism. Under the control of genes, our cells must multiply and transform themselves in very special ways before we can function as working life-forms.

BECOMING AN ORGANISM

Think about your home. It is made from different materials such as bricks, stone, concrete, wood, plaster, glass, tile, fabric, metal. It probably took many weeks to build and, if you include the time needed to man-ufacture all of the separate components, many months. It's not a trivial task, building a home from scratch.

Now imagine you need to build not one but thousands of homes. And power stations, electricity supplies, water and gas networks, roads, shops, and businesses. It's getting hard, isn't it? So many different designs, materials, and organizations that all have to be fitted together.

Now try to get them to build themselves. No people, just buildings that make themselves. Sounds impossible? Well, one final little problem for you: imagine that you must start with only a single brick. That brick must build all of the thousands of buildings, constructing and placing all the materials. And we'll give it, say, nine months.

Clearly you can't do it. Our technology is not up to the job. But nature's technology does the equivalent for every new creature born on this planet. And it doesn't always need nine months—it takes only three weeks to make a mouse.

GOING THROUGH A STAGE

Among many of the famous scientists at University College London, we're lucky enough to have Lewis Wolpert, perhaps the most distin-guished embryologist in the world. Lewis is quite a character. He is known to be an extreme reductionist, and his views are sometimes hard to agree with. But he certainly knows embryology. We spoke not long ago, him sounding surprisingly similar to the actor Tom Baker during an enthusiastic performance as Dr. Who—sometimes whispering, some-times speaking loudly and melodramatically: "There are four main processes in development: the first is spatial organization, or *pattern for-mation*. That is how different parts of the embryo acquire different iden-tities. Then you have *cell differentiation*. Because cells are in different positions, they acquire different characters and ultimately end up as mus-cle and cartilage and that sort of thing. Then you have *morphogenesis,* or change of form. And that's about the forces that mold the shape of the embryo. If pattern formation is like painting, morphogenesis is like sculpture. And then you have *growth*."

I couldn't have put it better myself (as I'm sure Lewis would enjoy telling you). Four processes, caused by the expression of genes within cells, to make every organism: everything from a mouse to us. At the risk of incurring more of the wrath of Lewis,* let's look at these processes in more detail.

CELLULAR PATTERN FORMATION

The zygote begins dividing shortly after fertilization. Before long, there is a cluster of cells growing ever more numerous. But how is this cellular clump going to become an organism?

The first thing a set of developing cells needs to know is which way up they are. A plant needs to make sure stems go up and roots go down. An animal needs to have a mouth at one end and an anus at the other. If the cells cannot distinguish one end of their developing clump from another, then there is no way for them to form the right organs in the right places.

Cells use a number of tricks to determine up and down, left and right, back and front. For amphibians, the point at which sperm enter the egg during fertilization helps to define the head-tail orientation. In chicks, gravity helps to tell cells which way is up, as they float on top of the yolk in their eggs. But for mammals such as the mouse, all directions are thought to be defined by cell interactions: protein messages sent between cells. It is possible that the cells of mammals "decide" which end should be the head by signals sent from the uterus as the clump of cells becomes implanted. As yet, we do not know for sure.

Nevertheless, the cells of all developing organisms will have worked out their orientation at a very early stage. And once they know their ups from their downs and lefts from their rights, they can work out exactly where they are within the multiplying group of cells.

To explain how, let me take you to a small town I know of. You have no idea what the place is like, and to make matters worse, I'm going to blindfold you. The only thing I'll tell you is that there's a busy railway station to the north of the town and a single large bakery to the east. Let's meet in the town square, which is at the center of the town.

I leave you alone, blind and confused. But you quickly discover that it's very noisy where you are. You hear the sounds of trains nearby. You

---

*But usually with a twinkle in his eye.

realize that you must be close to the station and so must be at the north end of the town. You start to walk away from the noise and eventually reach a point where you can hear nothing. You have walked to the south end of town. You turn around and try to walk until the noise of trains is just audible. And before long you start to smell fresh bread. You must now be at the east part of the town. So you turn your back on the baking smell and walk west, adjusting your direction to keep the sound of the trains on your right at a constant low drone. Before long, you arrive at a point where the baking smell behind you is faint and the noise of trains on your right is low. And you've found the town center and me— well done!

What you did was use the smells and sounds as a coordinate system. You were able to judge your location in the town based on the strength of the signals coming from the north and east. And as should be obvious by now, this is how our cells work out where they are in the growing clump. With protein signals being relayed from various key parts of the cluster, cells closer to a signal source receive higher concentrations of the protein message than cells farther away. In exactly the same way that we use three-dimensional coordinates to define the (x,y,z) positions of things in space, our cells produce (x,y,z) chemical gradients to let each other know where they are. And because different concentrations of proteins trigger different genes, our DNA is able to use this positional information to make cells at specific positions do specific things.

So protein messages are used to help create the first patterns of organization within developing cells. But some scientists go further, arguing that certain proteins chemically interact, producing patterns of chemical concentrations that are then used to help specify corresponding patterns of cell behaviors. As we saw in the Plants chapter, reaction-diffusion patterns may be responsible for patterns on snail shells, spots on giraffes, and stripes on zebras. But the evidence for these effects in our development is still low. Lewis Wolpert said on this subject, "I know the limb very well. I would like the digits to be a Turing pattern. It would suit me very well. I just don't know how to do the experiment. Are leopard spots caused by Turing patterns? The blunt fact is that I don't know."

But whether our development uses Turing patterns or not, we know that patterns of cell behavior are laid down according to their positions by protein interactions among multiplying cells.

DIFFERENTIATION

Cell behavior. What does that really mean? We've already had a good look at one important behavior that our cells have: producing proteins from genes. But cells do more than that. Neurons produce electricity. Muscle cells produce movement. Immune cells consume pathogens. Where do all of these specialized cells come from? After all, at first, there is only a cluster of similar-looking, nondescript cells.

It turns out that our 250-plus specialized cell types are created in a process known as *differentiation*. As the cells receive different protein combinations and concentrations according to their location, different genes are expressed within them. When cells with certain combinations of active genes divide, their children are subtly different. Aided by the protein messages from the cells in their environment, the children have a slightly different pattern of gene expression. So when they divide, their children are subtly different from them, and their children different again. Over many generations, cells become more and more specialized at different tasks. It's like a handyman having a son who becomes a carpenter, who has a son who becomes a cabinetmaker, who has a son who becomes an expert on marquetry for cabinet doors. Each child is more specialized than the previous one.

So as we grow, patterns of specialized cells slowly form. And because we end up as a collection of mostly specialized cells, millions must die to make this happen. Indeed, most cells seem programmed to die unless rescued by a signal to tell them that they are good enough to become part of the finished organism. However, some cells remain as general-purpose mass producers. Called stem cells, they live in places such as our bone marrow, producing more copies of themselves and all the cells that make up our blood (see the previous chapter).

Patterns of differentiating cells are not the only means by which life is created. To make an organism, cells must work together and form many different shapes. Eyes should be spherical (more or less), intestines should be tubular, kidneys should be kidney shaped. How can a ball of changing cells alter its own shape so extensively and precisely?

MORPHOGENESIS

You guessed it—cells form larger shapes by using yet another set of tricks, this time falling under the name *morphogenesis.*

No single cell knows anything about the whole organism. It has no idea whether it is destined to help shape a foot, a heart, or a brain. Cells just produce proteins according to their genes, which affect other cells in their immediate neighborhood. And yet when a developing embryo is studied, we see astonishing changes of form. Large sheets of cells fold themselves inward to form internal hollows and tubes. Corrugations form, vast groups of cells elongate, twist, and move. Structures seem to turn themselves inside out. Our genes take the art of origami to a new level as they help to build us. And it's all done with proteins.

The key abilities that cells use during morphogenesis are stickiness, change in shape, and movement. Let me first explain the sticky side of things. We've seen that cells produce different proteins as they differentiate. Well, proteins on the surface of cells can be quite sticky. Two cells of the same type often stick to each other, whereas cells that are different will not. For example, cells that form the neural tube (which eventually develops into brain and spinal cord) adhere to each other but not to the surrounding tissue (ectoderm).

Different degrees of stickiness (or adhesion, to use the proper terminology) are important for morphogenesis. If no cell adhered to any other, we could never become a multicellular organism. But if every cell adhered indiscriminately to all other cells, then we could never be anything more than a big round sticky ball of cells. By enabling different cells to clump together into distinct groups, our genes allow separate structures to form and be moved relative to each other. This is how our organs can be made without running into problems such as the heart getting stuck to the lungs or the liver adhering to a kidney.

So similar cells stick and different ones don't, allowing whole structures to be moved independently of each other. But how does nature ensure that when the separate developing structures are moved and folded, the folds will be in the right places? If you push two ends of a flat sheet of paper toward the middle, it may bend in many different ways. Nature must ensure that sheets of cells always bend and fold in the right way. As usual, the answer lies with proteins.

It turns out that cells have muscles. Within our cells are networks of filaments—microscopic wires made of a protein called actin. When a cell divides into two, it is split down the middle by the constriction of an actin filament, like a balloon squeezed by a tightening length of string wrapped around its middle. But actin is also used to reshape cells. By contracting filaments wrapped around one end, the cell can be forced into a wedge shape. Different filaments allow cells to be elongated, warped, and squashed.

When a fold needs to be positioned in the right place, cells are programmed to change their form. They turn themselves into lots of little wedge shapes, which together create a small dimple or crease in a sheet of similar cells. And once you have a crease, you know that when you push the groups of cells together, they will bend in the right place. The art of origami with proteins.

But sometimes even the most convoluted folding is not enough to get the right cells in the right places. On these occasions, cells are designed to move themselves along developing tissues. For example, once cells have been pushed into the neural tube, some migrate away, later helping to form nearby structures such as neurons, adrenal glands, and cartilage. They move by using a combination of stickiness and actin filaments. Like a truck winching itself through mud, the clever cells throw out parts of themselves containing bundles of actin filaments. These stick to cells in front of them, and they then contract the filaments, winching themselves forward. Once in position, the cells act as seeds, generating many new offspring that grow into organs and structures in those hard-to-reach places.

GROWTH

We've now reached the fourth and final major stage in development. Organisms are formed through pattern formation, differentiation, and morphogenesis, but it all happens at a tremendously small scale to begin with. Even when most of the major structures in a human embryo are formed (after eight to ten weeks), the total size of the embryo is little more than an inch. Before this tiny embryo can become a functioning human being, it must *grow*.

Clearly embryos grow larger throughout their development. It's a process that is needed to help push and mold organs into the right forms. But the vast majority of growth happens when embryonic development is largely complete. And growth happens using three mechanisms: cell proliferation, cell enlargement, and accretion.

The first sounds simple but is rather complex. For cells to proliferate, or multiply, they must divide. Each cell is triggered to duplicate itself by growth hormones—special chemicals produced in glands, which (for mammals) are under the control of the hypothalamus in the brain. Early cell division is automatic and does not involve any great increase in size. Dividing cells simply become smaller and smaller. But later, cells must increase their size before dividing. Not only that, but the whole process of division is very tricky. Exact copies of all of the DNA in the nucleus must be made and separated. The structures in the cell must be duplicated, and everything must be split down the middle (using actin fibers, as we heard earlier). If any mistakes are made anywhere, the cell will not work and may die. Even worse, if the DNA is not copied accurately, the new cell might start to proliferate uncontrollably, resulting in cancer.

Another method used to grow organisms is cell enlargement. Quite simply, it means the increase in size of individual cells, which obviously results in the expansion of the organism as a whole. Plants make use of cell enlargement quite a bit, as water is pumped into cells by osmosis, expanding them and helping to increase the speed of their race toward the sunlight.

Finally, there is accretion. Not every part of a multicellular organism is alive. Evolution discovered millennia ago that certain substances are harder and more protective than living cells could ever be. It learned that with the right combination of genes, cells could produce substances that solidified into hard, protective layers. And so we evolved to develop the hard internal scaffolding that surrounds our delicate organs and gives us extra mobility. This scaffolding, which we call bone, is formed by special cells that secrete chemicals, forming cartilage. The cartilage then slowly becomes converted into bone as lime salts are deposited by other cells (ossification). Other examples of accretion include the formation of cartilage in our ears and nose and the growth of our teeth.

Growth is a complicated process. It must be perfectly controlled and maintained through the remaining months before birth, and also for many years after birth, as the organism matures into adult form. Proportions must be adjusted, bones lengthened, lungs expanded, muscles increased. Like the processes of pattern formation, differentiation, and morphogenesis before it, growth is a fundamental reason why you are the way you are.

GROWING INSIDE COMPUTERS

In 1998, an archipelago of sixty Japanese islands supported an exotic form of primitive life. Not much different from bacteria, its populations would compete for food, reproduce, and evolve. But unlike bacteria, the creatures were multicellular. Each organism developed according to its genes from a single cell into an adult form comprising ten cells. Like most natural development processes, the cells divided and differentiated in order to form the finished organisms. Two cells were used for reproduction and eight for sensing their environment. Using their senses, the organisms were able to detect whether nearby islands had more food. If they saw that "the grass was greener," they would move to the better island. This was sometimes essential, for individual islands would undergo occasional environmental disasters that wiped out all life on them.

Like the evolution of all of nature's organisms, the form of these creatures had been generated through the evolution of their genes. The programs of development were written by a gradual process of selection, variability, and heredity, resulting in complex little life-forms that were designed to maximize their survival capabilities. But unlike in nature, the creatures had no physical form. Their islands had no water around them or soil or plants on them. These creatures were machine-code computer programs, and their islands were digital domains within a network of sixty computers.

Called Network Tierra, this was the pioneering work of biologist Tom Ray and his colleagues. They showed how digital organisms (their name) could evolve within the memories of computers. Instead of self-replicating molecules like DNA, the organisms used self-replicating strings of machine-code instructions. These formed miniature programs that produced new copies of their own code. Their food was computer processor cycles: the more they got, the faster they could function. But because of competition for this resource (they all want to have as much processor time as possible), they had to evolve strategies to outdo each other. Like the race of plants toward sunlight, Tierra's organisms had to evolve to become ever more efficient at snatching a few more nanoseconds of processor cycle.

The work of Tom and his team is intended to discover the limits of digital evolution within computers. Their aim is not to copy processes of development such as differentiation directly. Instead, they create digital environments in which processes such as evolution can occur. With minimal assistance from them, they wish to observe whether developmental processes spontaneously form. Like many other researchers (including myself), Tom Ray believes that a key feature of tomorrow's evolutionary algorithms should be development. And by demonstrating that developmental processes analogous to those that evolved in nature will also arise within computers, he shows us that evolutionary computation may have the same capabilities as natural evolution. It's also a lovely glimpse into our origins on Earth, for Tierra shows just how changes in genetic code can lead to multicellular organisms being built using processes such as differentiation.

Tierra has become famous since its beginnings in the early 1990s. It has illustrated exactly how self-reproducing code constantly improves during evolution. We have seen the creation of Tierra parasites that attach themselves to certain organisms and live off them. And the most recent work has shown how the machine-code equivalent of cells are used by organisms to improve their survival by differentiating and growing efficient sensors.

EXPRESSING DIGITAL GENES

One of the reasons Tierra is seen as such a success is that processes of development and gene expression are very new to most areas of computer science. As we saw in the chapter on evolution, evolutionary algorithms have none of the processes used in nature to create organisms from their genes. In a genetic algorithm, for example, a binary gene is converted straight to a parameter value. So traditional computer evolution really does use the equivalent of genes for "length of legs" or "eye color."

Not surprisingly, this has long been the source of much scorn from developmental biologists. In nature, genes must be expressed through a highly regulated network of protein interactions, which results in the development of organisms. Our genes define the recipe used to make us—they certainly do not define parameter values that specify our dimensions like a draftsman's plan.

So in recent years, researchers have begun to change the way evolution is performed within computers. The first important advances began with the invention of new types of genetic algorithms, such as the structural GA. Instead of using every gene to define a set of parameter values, which then defines a solution to a problem, structural GAs employ conditional genes. Like the "IF . . . THEN" structures in our own genes, these genetic algorithms use regulatory genes to trigger or suppress the activation of other genes in a digital chromosome. These ideas have been used with some success to evolve solutions for engineering design problems.

In the chapters on evolution and plants we saw other examples of complicated genotypes that define growth programs: genetic programming and Lindenmayer systems. Both permit genes to be triggered by other genes and be reused. But these methods are all still worlds apart from natural gene expression.

Researchers have literally just begun the investigation of biologically plausible gene expression algorithms within computers. A small but rapidly increasing number of researchers worldwide is beginning to show that digital genes that trigger or suppress each other—sometimes using analogs of proteins—may become a powerful way to increase the capabilities of evolution within computers. Some think that these ideas will help us to evolve better computer programs, others think that more advanced neural networks can be evolved. But we are now standing at the forefront of science. The field of computer gene expression is so young that its age is still better measured in months rather than years.

Until it has had more time to mature, we will not be sure what techno-logical advances digital gene expression will enable.

CELLULAR AUTOMATA

Not all computer-based developmental processes are so new. *Cellular automata* were the brainchild of John von Neumann (whose von Neu-mann architecture became the basis for all modern computers, as we saw in the Universe chapter).

Around fifty years ago, Johnny (as he was known by all) was inter-ested in self-reproduction. Using pencil and paper (for there were no computers powerful enough to automate the process), he designed a different kind of computer. The cellular automation (or CA to its friends) was the result—a clever idea that somehow managed to become embedded in computer science, artificial intelligence, and biol-ogy folklore for the next fifty years. As I've mentioned in many of the previous chapters, CAs are now used to model crystal growth, reaction-diffusion systems, population dynamics in ecology, immune systems, and insect swarms. They've also been used to describe traffic move-ment, urban growth, and patterns of spreading fire and to generate music. But most significant here, CAs have been used to model the development of organisms.

It was twenty years after von Neumann invented the CA, shortly after Marvin Minsky's work on CAs, and shortly before Stuart Kauffman began his research on the topic. Cambridge mathematician John Horton Conway was placing dishes on the checkered floor of his hallway at home. Starting with a few randomly placed dishes, he would add and remove dishes according to three simple rules:

The Rule of Birth
If exactly three tiles on the floor have dishes on them and are neigh-bors of a tile without a dish on it, add a dish to the tile without one.

The Rule of Overpopulation
If a dish has four or more dishes on the tiles next to it, remove the dish.

The Rule of Underpopulation
If a dish has one or no dishes on tiles next to it, remove the dish.

As well as producing large amounts of broken crockery, John Conway had devised the most notorious CA of all. Today known as Conway's "Game of Life," it shows how three simple rules can work together to produce unlimited complexity. Beginning with a few randomly placed dishes, astonishing patterns develop, expanding, contracting, and sometimes moving across the tiled floor (see plate 15, top). These patterns are the result of the same simple rules being reapplied over and over again. The development of these hypnotizing forms has been likened to the development of primitive life in nature.

Over the years, thousands of eager programmers have reimplemented the program and spent many hours watching the never-ending patterns that form. Of course, these days we do not use plates and tiled floors. We use digital universes in which little dots on the screen represent the dishes, with computers carrying out the rules.

The patterns of dots produced by Conway's Game of Life have been studied so extensively that there is a whole list of names for the different forms that arise. For example, the "block" and "beehive" are static patterns that do not change. "Blinkers," such as the "traffic light," alternate between two or more patterns. One of the most famous is the "glider"—a pattern that repeats itself every four iterations, except that it has moved on the "tiled floor" (grid). Gliders will often be seen walking across computer screens until they collide with other patterns, causing new patterns to form. People have also carefully designed forms that automatically produce gliders; examples include the "glider gun" and "puffer train."*

Today there are countless different types of cellular automata. Some live in one-dimensional universes (and have been used to perform mathematical calculations via their interactions). Others, like the Game of Life, live in two-dimensional universes. Still others live in three-dimensional universes, creating three-dimensional forms through the interaction of their rules. Each different CA has a different set of rules and a different number of states (so instead of "dish" or "no dish," some might have "no

---

*If you want to play with a CA yourself, do a search on the Internet for "cellular automata." You'll find many Web pages that contain living CAs that produce patterns before your eyes.

dish" or dishes of ten different colors; see plate 15, bottom). And some CAs are capable of supporting patterns that can reproduce themselves.

But what has all this to do with development? Recall that our own development is the result of a self-regulating network of interactions among cells. Genes produce proteins that trigger or suppress genes in other cells that produce proteins. The signals cause cells to multiply and form patterns.

Cellular automata are surprisingly similar in many ways. Patterns are produced by the interaction of cells (Conway's dishes) according to a set of rules. Instead of using proteins to activate genetic rules, CAs use the position and status of cells to activate their CA rules.* If you're in the right place, you can trigger the addition or removal of "dishes on the floor." So every cell of a cellular automaton affects the status of neighboring cells, which affect other cells, and so on. In addition, like genetic instructions within biological cells, CA cells all "contain" the same set of CA rules. And in the same way that each biological cell will have only some of its genes expressed, most CA cells will express only some of the CA rules at any one point in time.

With the right rules and the right starting pattern, a CA will develop specific forms inside a computer. And researchers are beginning to make use of these developmental abilities. Scientists (most notably Hugo de Garis, whom we met in the Brains chapter) are using evolution to help generate special CA rules. The resulting CAs then develop neural networks, helping to specify how digital neurons should be wired and what they should do. Others evolve CAs to help generate novel electronic circuits or architectural designs. And as we saw earlier, CAs are used to model the behavior of just about every dynamical system conceivable. Cellular automata have long been the mainstay of digital development.

The abilities of cellular automata to duplicate the behavior of so many biological processes have led some researchers, most notably Stuart Kauffman, to develop a theory. If CAs are so good at modeling different

---

*In CA parlance, the word *cell* means "grid position." So a cellular automaton is a grid of cells, which may be empty or filled (or have other states for more complicated CAs). For example, each of John Conway's floor tiles was a cell, which could be empty or could contain a dish.

dynamical systems, perhaps some process within them can also be found within all those dynamical systems. Perhaps at some level of abstraction, immune systems, brains, growth, insect swarms, and cellular automata all look very similar. If true, then the abstracted model should explain and predict some of the monstrously complex behavior of these diverse systems. This was Stuart Kauffman's thinking as he created his model of *Boolean networks.*

BOOLEAN NETWORKS

The word *Boolean* means two-valued: true or false, 1 or 0. Kauffman's Boolean networks are generalizations of CAs, where every node in the network has a Boolean value. The networks are more flexible than standard CAs, for it is possible to connect any node to any other node (each CA cell is connected only to its immediate neighbors). In *At Home in the Universe,* Stuart gives an illustrative example of a Boolean network: a network of lights, each one able to shine red or green. The color of each light is set by a collection of rules based on the connections between the lights. So if a light is connected to five other lights, one of its rules might be: "If the first three lights are green and the last two are red, then shine red." Another might be: "If the last light is green and I'm shining red, shine green."

Clearly, this is much the same as a cellular automaton with its rules that change the state of cells. Instead of dishes on tiled floors, we have red/green lights wired together, but the way both systems operate is the same. Both use rules to change the state of cells based on the state of others. And within Boolean networks, patterns of different "colored lights" form, just as patterns of dishes formed on John Conway's floor. But Kauffman goes further. He argues that Boolean networks are also similar to neural networks, immune systems, and networks of gene expression. Not only that, but he has spent over thirty years studying the properties of Boolean networks in an attempt to understand these more complex biological systems. He has discovered that networks with many connections between each node will rarely settle into a stable state; the red and green lights will twinkle unceasingly. But networks in which each node connects to only two other nodes will often settle in stable patterns: large areas of static red or green lights form. Through years of such experiments, Kauffman believes that the kind of dynamic patterns that emerge in biological systems require networks that are somewhere between the two—on the boundary of chaos and order.

In recent years, Boolean networks have been extended to enable nodes to have more than two states. But opponents still argue that Kauffman's ideas are so abstracted from reality that his conclusions tell us nothing about the workings of biology. I asked Steve Jones for his response to Stuart Kauffman's name: "I don't understand him, is the answer. I class myself—always—as a field biologist, really. Where's the experiment? . . . I just don't see that it's telling me anything, but it may be that I'm too stupid to follow it. In fact, that's quite possibly true."

I also asked Lewis Wolpert about Stuart: "A great personal friend," he replied.

And his ideas?

"Totally irrelevant. You find me a developmental biologist who takes Kauffman seriously, and I'd be very surprised."

It seems that while Kauffman's ideas may have a strong following in mathematics and computer science, the biologists will take more convincing. And who is to say that they're wrong? It is quite possible that Boolean networks may be an example of a dynamical system native to our digital universes. Although they have their own kind of adaptability and diversity of behavior, they might not be able to predict or explain the processes of development to us any more than a swarm of insects could.

DEVELOPMENT WITH DIGITAL CELLS

So if cellular automata and Boolean networks aren't enough, what next? Recent research, led by computer scientists who want to increase the capabilities of evolutionary computation and biologists who want to model development, is looking elsewhere.

As we saw earlier, our development is caused by the impressively complex abilities of cells, as genes produce proteins that express the genes of other cells. Without clever cells, we could not develop. So researchers are now attempting to create clever digital cells that will enable more realistic and useful digital development.

One such approach has been created by Moshe Sipper and his team at the Swiss Federal Institute of Technology. They call the idea embryonic electronics. Rather like a CA, they form a grid of cells or "biodules." These are physical devices—cubes (with little computers inside them) that can be stuck together like oversized LEGO bricks. Instead of each biodule having a simple state such as filled or empty like a normal CA, each can contain a "genetic program" or be empty.

Development begins with a grid of empty biodules. A mother cell is placed into one biodule, which then replicates its genetic material, filling neighboring biodules. These also replicate, filling more biodules and eventually growing to fill the grid (if that is what their genes tell them to do). As they grow, different internal genes are expressed or suppressed based on their relative position from the mother cell. This causes "living" biodules to differentiate and enables an entire electronic organism to develop in the grid, with each biodule containing one of its electronic cells.

The idea allows electronic hardware to develop its own function automatically by following the developmental program in digital genes. It also enables some other exotic features: if there are sufficient biodules available, the collection of cell programs inhabiting them can reproduce and generate a spare copy. In other words, the electronic organism can have children within the grid of biodules. Alternatively, if some of the biodules are damaged, resulting in cell death of part of the organism, new cells automatically grow into nearby spare biodules, restoring the digital beastie. The ability of circuits to reconfigure and regenerate themselves if damaged would be of enormous benefit to the space industry for satellite and spacecraft electronics.

Other researchers are developing electronics in slightly different ways. Genetic programming guru John Koza's approach is to evolve developmental programs called cellular encoding that build analog electronic circuits by adding to an initial "embryonic circuit." He and his team have had considerable success in using their computer to discover many useful circuit designs.

But most of the work in this small but expanding field concerns the evolution of digital genes that define the development of neural networks, which then control robots. Or to put it another way, these guys are trying to grow robot brains.

As we saw in the chapter on brains, recurrent neural networks are very dynamic and adaptive digital brains, able to control robots rather well. But because we're not entirely clear on how they work, we're also not too clear about how to make them. How many neurons, which should be connected together, what the input weights should be: we don't know the answers. So the best solution is often to let evolution generate these digital brains for us. Unfortunately, as we become more ambitious with our robots, we need to generate more and more complex digital brains. Many researchers now believe that traditional evolutionary algorithms are simply not up to the job, so research into creating true digital development of neural nets is becoming popular.

The department of Cognitive and Computing Science at Sussex University leads the world in this area. Inman Harvey, Nick Jakobi, Torsten Reil, Frédéric Gruau, and others have all produced various developmental techniques.* Their work often employs digital chromosomes with many of the same features as our own DNA. Digital genes are expressed or suppressed, producing the digital equivalent of proteins that diffuse and interact with genes in other growing neurons. Using these processes, neural networks are developed from digital genes, axons and dendrites are formed, and differentiation into different types of neuron occurs. Although still in the "blue-sky" stages of research, such work is beginning to show real promise for the creation of complex brains within computers.

MOTIVATIONS AND ASPIRATIONS

Now we get to the point of computational development. And there is a point. We're enthusiastic about adding developmental processes to evolutionary algorithms not just for the challenge of doing it (although it is a considerable challenge). Development is actually a very useful process for computers.

One of the cleverest things about it is *scalability*. When we use traditional genetic encodings within our evolutionary algorithms (where one gene develops into one parameter of a solution), we run into trouble when we need complicated solutions. The more complex a solution to a

---

*But then, so have other researchers such as Frank Dellaert and Randall Beer of the Case Western Reserve University in the United States, and Jari Vaario of Nara Women's University, Japan.

problem is, the more parameters it needs, and hence the more genes we need to evolve. Not only that, but really complex solutions tend to have all kinds of organizations in them. A hospital design, for example, is made from a collection of similar buildings that contain collections of similar rooms and corridors, which contain collections of similar equipment and desks. Using our traditional evolutionary approaches to generate such hierarchical complexity is very hard indeed. And it's all because we don't use growth processes within computers.

We have just discovered that when we use computers to evolve the right kind of "development programs," all of these problems seem to be solved. Because evolution generates a set of instructions and not a solution directly, you don't need more genes in order to get more complicated solutions. In nature, the number of cells in an organism is unrelated to its total number of genes (a blue whale has much the same number of genes as a mouse, but it has trillions more cells). In exactly the same way, the number of digital genes does not need to increase as the size of the solution increases. So evolution should be able to generate designs of entire hospitals as easily as it can generate the design of one room—if we use development.

Not only that, but processes of development are rather good at producing solutions that display certain kinds of order in them. Because the same genetic instructions will often be reused quite a few times, certain patterns such as repetition, segmentation, symmetry, and subroutines will tend to form naturally. The same kinds of hierarchical complexity in complicated solutions that are so hard for traditional evolutionary algorithms to discover suddenly become easy to find when we evolve developmental processes.

Development has some clever abilities: scalability, complexity, and others we have already seen, such as self-regulation and self-repair. These provide a strong motivation for us in computer science to understand and use development within our computers. But doing so is very hard. Some of the best results so far were produced by Peter Eggenberger at the University of Zurich. He created an evolutionary system that made use of gene expression and regulation, positional information by gradients of proteins, cell differentiation, division, and simple morphogenesis. But even this work managed to produce only very simple solid, symmetrical multicellular shapes. It's all a long way from using these methods for solving our problems.

Nevertheless, my own view is that evolution without development will ultimately prove too limited. So work at UCL performed by my doctoral student Sanjeev Kumar and me, with the support of Lewis Wolpert (as well as researchers in many other university and industry groups), is trying to understand and use development for our future technology. Early experiments already seem to show massive scalability of the approach. We've all barely begun, but with a bit of persistence and luck, we should have the next generation of digital biology maturing in ten years or so.

DNA COMPUTING

If we can use digital DNA within our computers to develop solutions to problems, what about the other way around? Could we use biological DNA as computers? Could biological computers *made out of DNA* solve problems for us?

It sounds a little overambitious even for a science-fiction novel, but the answer to these questions is yes. As you read this, researchers are now making DNA computers. These computers have already solved simple traveling salesman problems (like those we saw in the Insects chapter). It has also been suggested that they may be able to crack the "unbreakable" encryption codes like those used by the U.S. government's Data Encryption Standard.

DNA computers are possible only because of the impressive array of techniques we now possess to manipulate, read, and create DNA molecules. They work in a rather different way from any other computer in the world. Instead of storing information as electronic 1s and 0s within silicon chips, DNA computers actually use synthetic strands of DNA to store information.

The idea makes a great deal of sense when you think about it. We know that our own DNA stores many megabytes of information within our forty-six chromosomes. We also know that DNA is very, very small. So small that even a test tube could contain trillions of different DNA molecules. And that's a lot of storage space.

This becomes significant when we consider certain types of problem. The traveling salesman problem, for example, involves finding the shortest route between a number of cities while visiting all of those cities. When you have many cities, there can be millions or even trillions of possible routes. Finding the right one can be very difficult. Alternatively, if you're trying to crack the encryption on a file, you need to discover the unique key that unlocks the file. Most of today's encryption methods have vast numbers of possible keys (for example, each file encrypted according to the U.S. Data Encryption Standard uses one of 73 quadrillion possible keys). Computer encryption is a few billion times more secure than the lock on your front door.

If you were to use a traditional computer to search through all of the possible solutions to problems such as these, it would take hundreds or thousands of years, even if you used the best supercomputers. But DNA computers are different. Instead of generating one solution at a time and checking it, a DNA computer would generate trillions of solutions at a time, each one defined by a slightly different strand of DNA. It might be possible to generate DNA molecules (either in a test tube or attached to a gold-coated square of glass*) that correspond to every possible solution to the problem. But how do you find the single molecule that defines the right answer?

This is the clever bit. Special enzymes are introduced, which alter the DNA molecules according to their pattern of bases. So, for example, enzymes might add a short base sequence if two specific patterns occurred somewhere in the DNA molecule. By using many of these operations, the enzymes can be used to perform logical and mathematical functions, just like the ALU of an ordinary computer. They can also be used to destroy any DNA molecules that represent invalid solutions. And because the enzymes will modify all of the billions of DNA molecules in parallel, the DNA computer effectively performs billions of computations at the same time. When the chemical reactions are finished, the best solutions can be picked by extracting the remaining DNA

---

*These are the current favored approaches by teams at the University of Wisconsin, Madison, and the University of Southern California.

strands. These can then be decoded using the same DNA fingerprinting techniques used by police to identify criminals.

The idea is remarkable, and for these special kinds of combinatorial problems, DNA computers could become unbeatable. Some experts also believe that DNA computers would be the obvious solution for the storage and retrieval of human DNA. If governments one day decide to hold records of every person's DNA code, they'd need millions of high-capacity CD-ROMs. Using DNA computers, the information contained in the DNA of every human being on the planet could all be held in one test tube.

But we still have a very long way to go before these biological computers become viable. Right now we do not have the technology to automate the large-scale synthesis of specific DNA molecules. The process of decoding the DNA is also still too slow. We may have to wait another twenty years before the first DNA computers are used to solve genuinely difficult problems. And it may be more than fifty years before you have a desktop DNA computer. But it will probably happen.

SUMMARY

Nobody ever claimed that growing up was easy. And the processes of development are certainly not simple. Evolution made our DNA, and our DNA made us. Coiled up within our cells, fragments of this clever molecule are transcribed into RNA and translated into proteins. Different genes produce different proteins, which produce different cells. As the genes and proteins regulate each other, so our cells divide and change. First, patterns of cells form, using protein coordinate systems. Then cells differentiate into different types and stick to each other, fold, and move. And of course they grow, multiplying, expanding, and excreting substances.

Development is still a new idea within our computers. But little digital multicellular organisms do grow in their digital universes. Some are machine-code programs, fighting for processor cycles. Others are cellular automata. And the processes of gene expression and development at last are emerging to take their place alongside evolution in their efforts to solve our hardest problems. From self-repairing electronics to the development of robot brains, this looks set to be the way forward for digital biology. Maybe one day, digital organisms will grow in digital universes created by DNA computers.

These marvels of nature and technology are only just beginning. As we understand more of the world around us, our technological capabilities increase. And as we create more digital biology, so our understanding of the world increases. Fueled by these leaps of knowledge, scientists such as Stuart Kauffman have discovered that much of biology seems to rely on surprisingly similar processes. Although his Boolean networks may not explain everything, he does have a point. When you examine a number of important aspects of nature, as we have done in the previous seven chapters, you can't help but notice a few common ideas. In the last chapter of this book, we'll see what it all means.

We'll start by watching some children. It's amazing what you can learn from them . . .

ANSWERS

"THERE!" CRIED RICK, pointing at a large tree on the other side of the field. After running all the way down the overgrown lane to their newly discovered patch of wasteland, the eleven-year-old was panting heavily. He looked around him and saw that the other four hadn't caught up yet. "Come on, come on!" he yelled impatiently. He pushed grubby hands through his dirty brown hair and frowned at them, wrinkling a freckled, stubby nose.

"Now what's he going on about?" Ben asked Sandy, as they jogged toward Rick, kicking autumn leaves to make them swirl around them.

"I think it's for the clubhouse," she replied, brushing a leaf from her usual oversized T-shirt.

Rick was now jumping up and down in front of them. "Oh, you're all so slow!" he exclaimed, as first Ben and Sandy, then Pete and Bryony arrived. "See that tree?" he said to them all, pointing again to the tree in the distance. It stood tall and defiant among the scrubland, its leaves a bright splash of green against the pale blue canvas of the sky.

"Yeah," came the simultaneous response from four mouths.

"It's puurrrrfeck! Don't you think? Don't you think?"

Rick was clearly getting overexcited, but Ben could see what he was talking about. It was a huge old oak tree that spread its boughs flat and wide, creating a large platform six or seven feet from the ground. It would indeed be perfect for the Foxes' clubhouse. "Ricky, baby," drawled Ben in one of his unidentifiable fake accents, "you done good, boy."

Ten minutes later, the distant sounds of "Foxes rule! Foxes rule!"

could be heard across the field. Once again in its hundreds of years of life, the old oak tree was host to a rowdy group of children.

"Okay, okay," said Ben, waving his hands to quiet his friends down. His dark brown eyes surveyed his friends, sitting around him among the large branches of the tree. "Club meeting is started. And I think we need a new Foxes' club rule."

"Yeah, like what?" said Pete, his feet dangling either side of a huge branch thicker than his chest.

"Like: nobody except Foxes is allowed in our clubhouse!"

"Yeah!"

"Excellent!"

"We still keep the other two, right?" asked Pete nervously. Pete was rather proud that he had thought of the existing rules, even if they had been heavily inspired by his older brother's gang.

"Course," replied Ben. "Rule number one: Foxes are better than anyone. Rule number two: Foxes help Foxes first. And new rule number three: Only Foxes in the clubhouse."

The only possible answer to such insightful rules was the inevitable chanting from the club members. "Foxes rule! Foxes rule!"

Over the next few days of their vacation, the gang of boys and girls dragged, hoisted, and tied innumerable pieces of wood and fragments of fence to make a platform in their tree. A makeshift roof was made from plastic sheeting to fill in gaps of the canopy above. They even constructed a lookout tower complete with telescope (Ben's sister's old one, secretly swiped). The group worked tirelessly, planning, cooperating, and coordinating its efforts in the intense way of children linked by a common cause. Back home, more than one of them now had slightly less sturdy fences around their back gardens and the prospect of angry parents in the near future. Parents who would not be able to understand why their children should invest so much time and effort into building a treehouse, when homework and chores were so neglected. But it would be worth it, for the clubhouse now had a more or less flat floor, seats, a waterproof roof, a retractable rope ladder, and even a carefully decorated "club box" painted by the artistic Bryony.

But the work of the Foxes had not gone unnoticed.

• • •

It was nine in the evening; the light was turning golden and the shadows getting longer. The Foxes were in their clubhouse playing a tournament on their pocket electronic games, while eating snacks raided from home. Bryony and Pete were in hysterics over a terrible joke, their laughter drifting across the quiet field.

The dappled light flickered on a round, pale object as it arced up through the lower branches of the tree, toward the Foxes.

"Wah! What was that?" cried Bryony, as she was struck on the head with something. Feeling with her hands, she discovered a broken eggshell and sticky, foul-smelling egg all over her hair. "Urrghh! Gross!" she shouted, startling the others.

"Hee hee hee," giggled Pete, "a bird is bombing Bryony! Hee hee— aaahh!" Pete nearly fell out of the tree with surprise as another egg hit him in the face.

"It's not birds," yelled Rick. "It's the Tigers!"

"That's right," came a threatening snarl from below, "and you're in our new clubhouse."

"No way!" cried Sandy. "We made it. It's *ours*."

"Not for long, suckers," came another voice, quickly followed by a rain of rotten eggs.

The battle that ensued was vicious but short. The five members of the rival Tigers gang hadn't expected the Foxes to stay put, so their supply of rotten eggs quickly ran out. The fight soon became a battle of words, each side trying to outdo each other's insults and volume.

This continued until Pete gave a yell. "Hey! Someone's up here! Ooof!"

"Quick," cried Ben, "he's on top of Pete! Kick him out!"

The other Foxes rushed to aid their fellow member and forcibly eject the invader. With a little more pushing than was necessary, the unlucky Tiger fell from the tree and landed on his arm with a shriek of pain. Meanwhile, Pete sat up and looked around in the clubhouse, a little dazed.

"Now you've done it," came a cry from below. "You've busted his arm. You're in trouble now." And the Tiger gang limped away, supporting its injured member.

"Did that kid really break his arm?" asked Bryony.

"Nah," replied Rick scornfully, "look at him running home now." As they watched, the child ran toward the lane, hurt arm forgotten, chased by his fellow Tigers.

"Are you okay?" Sandy asked Pete.

"Yeah, I just got kicked in the stomach, is all. Thanks for helping, guys."

"That's what we do," said Ben. "We're Foxes, aren't we?"

"Foxes rule! Foxes rule!" came the victorious response.

COMPLEX SOCIETIES

Many of us had these kinds of experiences when we were young: a group of friends united by common activities and beliefs. Such collections of people can have team spirit, bringing out the very best of humanity as the group achieves so much more than individuals could.*

Whether a family, a gang, a team, a community, a race, or a country, if you group people together, then the same things tend to happen. Individuals organize themselves, communicate, and interact like cells of a larger organism. Somehow the group takes on a life of its own, with its own needs and behaviors. The name, reputation, and territory of the group can become more important than the needs or desires of individuals. Through rules of behavior for its individuals, the group perpetuates and strengthens itself, fighting other groups for survival.

The fictional group of youths that we met here showed just a few of these behaviors. The Foxes developed a hierarchy, allocated different jobs to each other, and cooperated in order to achieve common goals. Because of their club rules, the children developed a special bond—a deep-seated pride in being a member of the club. This pride brought them together, unifying them into a cohesive whole. They became rather like an independent organism, which found itself a territory and built itself a home. This organism also defended itself and its home, improving its reputation in the process. When part of the club was hurt, other parts protected it and helped it to heal.

---

*But there's always the flip side: at their worst, such gatherings can deteriorate into gang warfare, bringing out the very worst of humanity and doing far more harm than an individual could.

The gang was made from the rules: "We're superior. We help each other first. We don't let anyone else in our home." As long as the rules were followed, then unity, group decision making, and territorialism were inevitable. And although these childish rules may seem extreme, they lie at the heart of many human groups. It is unspoken rules such as these, embedded within the cultures and upbringing of people, that have caused devastating wars. But it is also rules such as these that enable us to build societies, work together, and, through competition, advance our civilizations and technologies. Properly used, national pride can unite millions of individuals into a cohesive unit.

So what has this to do with processes of natural or digital biology? In fact, it has quite a bit to do with them. Groups of people tend to behave according to certain rules. The rules may be rules of a football game, rules of etiquette at a dinner, or laws and culture of a society, but whatever they are, the rules force individuals to act in certain ways. The result is group behavior: a complex overall behavior that may be quite different from that of the participants of the group. The individual may have only tightened a few nuts and bolts, but the group may have built a spacecraft. The individual may only have kicked a ball a long way, but the team may have won a tournament. A single person cannot understand everything, but our collective minds can understand amazing breadths and depths of our universe.

The capabilities of each of us are not significantly different from those of Stone Age humans, but together we have become gods, flying higher than any bird, swimming deeper than any fish, capable of moving mountains and dividing seas. We have altered the face of this planet, modified its life, changed its air. As a group, humankind is an unstoppable force—a super-intelligent, massively powerful entity.

And this entity is made from a collection of simpler things that follow a set of rules. Us.

RULES RULE

Like an ant colony or insect swarm, our groups and societies act as superorganisms, constructed from simpler things (people) that follow rules. Sound familiar? It should by now. Remember that each of us has a brain: a collection of simpler things (neurons) that follow rules. We also have an immune system, which is made from a collection of simpler things (immune cells and chemicals) that follow rules. We developed from a collection of simpler things (genes that produce proteins and cell behaviors) that follow rules. We're surrounded by plants that grow with astonishing patterns, fractals, and chaos caused by rules. Our genes and those of every other living thing were evolved through the action of rules. Everything is made from simple points of energy, existing within a universe that is made from rules.

There does seem to be a coherent theme in all of these natural systems. Lots of simpler things that obey rules somehow make all types of complex adaptive systems. Universes, evolution, brains, insects, plants, immune systems, and growth. They're all the same. They're all made from simpler things that obey rules. Those simpler things become more than the sum of their components. Complexity emerges from simplicity.

Our societies are not just groups of separate people. They are cooperating, planning, unified entities. Evolution isn't just a bunch of creatures reproducing and dying. It's an amazingly creative process that has enabled the most miraculous of forms to be generated. Your brain isn't just a collection of wineglass-like neurons—it's a conscious, thinking mind. An ant nest isn't just a swarm of creepy crawlies—it's a planning, optimizing intelligence. The patterns of plants aren't random—they're the result of golden mathematical laws selected for their near perfection. Your immune system isn't just a sloshing bucket of disparate cells and chemicals—it's a dynamic, adaptive, evolving maintainer of health. Your development from a single cell to adult form didn't happen because of just a few cells multiplying—it's an unimaginably complex network of gene, protein, and cell regulation.

Somehow, when lots of simple things get together according to a few rules, you get more out than you put in. And this is just not the kind of thing we're used to. In our everyday worlds, if you add one and one, you get two. But for these natural systems, adding one and one seems to give us ten. It's like employing some painters to redecorate one room and finding they've done the whole house in less time and at no extra cost—these things don't usually happen. And when we try to model and predict such systems with mathematics, we find that it can't cope; the math falls over. We cannot explain or predict how any of the biological systems described in this book really works with traditional methods.

But computers give us a little more. When we create digital versions of the biological systems, we are able to watch and analyze exactly what happens. We learn how to use digital biology within our technology, but we also learn about the biological processes that we are using.

It has become very noticeable to me, both through my research and as I wrote this book, just how much we have now found out about the complex adaptive systems of biology. It is also noticeable how isolated these understandings are. Each field, be it neuroscience or evolutionary computation, has arrived at its own theories of how a system works. Each uses different ideas, different terminology, and different viewpoints. But the thing that blew me away was how *similar* all of these different ideas are. Without knowing it, hundreds of scientists in different fields all seem to be saying the same things. They're just speaking in different languages.

MAKING COMPLEXITY

When you amalgamate the hundreds of similar ideas created independently by many different biologists, chemists, physicists, philosophers, computer scientists, and mathematicians, you find a law. It's a fundamental law of nature, life, intelligence, and maybe everything. And I think it's elegantly simple.

I think the law that enables our universe and all of the complicated things in it to exist is something like this:

Many interacting things create complexity.

Pretty easy, isn't it? But what does it mean? And does it even tell us anything useful?

It's a two-part law. First, you need lots of things that have behaviors. Second, you need rules that say the behavior of each thing must be determined at least partially by some of the other things. In other words, you need lots of interacting things. And with the right kinds of interactions, you get more out than you put in. Complexity *emerges*.

We can test this principle within digital universes where we turn off some of the rules. When only one thing exists, there's not much of interest. When things can't affect each other, it's no more interesting than a single thing. But when there are lots of things that can affect each other, stuff happens. Interesting stuff. Complex stuff.

But scientists have learned that not all types of interaction will give us this higher-level complexity. For example, think about a set of dominoes standing in a line. Push the first, and they all fall over in a cascade. We've got lots of things that interact according to rules. But we haven't really got anything more complex emerging here. You can replace twenty dominoes with fifty, or with two, and you'll see the same overall behavior: they all fall over. There's little higher-level order or pattern of behavior emerging here. In contrast, think about a group of people. They interact and communicate, altering each other's behavior. Higher-level group behaviors—perhaps the activities of a club or pattern of play of a football team—emerge as a result of the actions of individuals. You can't replace a group of five people with a group of fifty, or of two, and expect it to behave in the same way because it won't.

So what's the difference between a group of people and a line of dominoes? As before, we've got lots of things (or people) that interact according to rules. This time, we've got *feedback*.

It seems that our multiple things must change each other's behavior in feedback loops if we want complexity to emerge. If I push you, you must push me back, causing me to push you again, which causes you to push me back again, and so on. These feedback loops allow complexity to grow. There is no longer a direct, or *linear,* relationship between individuals. Instead, the behavior of one individual can become amplified until it affects many others, including itself. So instead of the linear relationship we see with falling dominoes, we need the nonlinear relationships of groups of people. One domino can only ever cause the next to topple and nothing more, but one person can affect many others, including himself or herself, in a dynamic, repetitive, and expanding ripple of interactions.

So we can now fine-tune our earlier law:

Many things that interact with feedback create complexity.

And yet this is still not quite sufficient. If you remember from the Universes chapter, in order to explain the creation of our universe, we need a rule that must be defined in terms of another universe. We need something external to our universe to explain its cause. Similarly, a football team needs external things—a football and another team—before it can do anything interesting. And a brain needs something external to itself before it can think. An interacting network of neurons can't learn or think about anything without some external thing impinging on the neurons. It's another feature common to all of these kinds of systems: there needs to be some kind of external disturbance to the system to trigger interesting behavior. So we can now fine-tune our law one final time:

> Many things that interact with feedback and are perturbed create complexity.★

With this simple law, we can understand how complex systems work. We can also see just how similar all of these systems are. And they are spookily similar.

EVERYTHING THE SAME

The Foxes club that we began the chapter with was a complex system made from five children. They interacted and worked together, achieving goals such as the construction and defense of a clubhouse. If you recall, they followed certain club rules, which dictated their behavior: "Rule number one: Foxes are better than anyone. Rule number two: Foxes help Foxes first. And rule number three: Only Foxes in the clubhouse."

By helping each other, the Foxes interacted and coordinated their efforts as a team. The better they worked as a group, the more they regarded themselves as better than others, such as members of rival gangs (rule 2 intensifying rule 1). Regarding themselves as superior to others

---

★Of course, multiple interacting things with feedback and perturbations may not always create *useful* complexity; they may get stuck in a stable state or fly off into chaotic randomness. But if you've got complexity—particularly the kinds of complexity we've seen in this book—you can be sure that it's because of our law.

helped to reinforce the rule to keep others out of the clubhouse (rule 1 intensifying rule 3). And by keeping others away, their ability to help each other and work together improved (rule 3 intensifying rule 2). It's an example of feedback caused by their rules, on top of the normal feedback caused by the interaction of friends.

Now and again, the Foxes were perturbed by external factors: the discovery of a tree triggered the clubhouse building behavior; the attack by a rival gang triggered the clubhouse defense behavior. The children interacted with feedback and the occasional perturbation, resulting in complex group behavior.

This is not the only kind of human complex adaptive system. The world economy is a dynamic and uncertain system. Multiple businesses, industries, and traders interact with feedback, as well as being influenced by external events such as natural disasters, scarcity of resources, or even just rumors. The result is a highly complex and unpredictable high-level behavior.

But we have already seen much better examples than these of complexity emerging in this way. Indeed, *every* major biological and digital system described in this book works in this way. From universes to molecules of DNA, all emergent behavior follows our simple law. To see this, let's look back at some of the natural and digital biology that this book has explored.

UNIVERSES

Universes are made from rules. The rules define how the particles (which are all points of energy) should interact. There's a great deal of feedback in the interaction of particles in our universe: particles push other particles, and they push back. We even have Newton's third law of motion: For every action, there is an equal and opposite reaction. With so many actions and reactions, pushing and being pushed, feedback is unavoidable. And in order to create the universe, we needed some external rule or perturbation.

Computer programs also turned out to be like this. Made from rules and executed on hardware that works according to electronic rules, they define how the particles (binary 1s and 0s) should interact. There's a lot of feedback in those interactions, as program loops and subroutines repeat and amplify the actions of other parts of the code. And to make programs start and do useful things, we need external perturbations—input by the user or other programs.

EVOLUTION

Evolution happens when populations reproduce with heritability, variation, and selection. These processes define how the particles, or genes, propagate within populations. There is interaction among genes—this time, indirect interaction—as the creatures that develop from the genes fight for survival. Feedback happens when reproduction and selection combine, producing ever more copies of better genes in the population and fewer copies of worse ones. And perturbations during the copying of genes cause mutation—the variation that provides the genetic novelty essential for the development of new species.

Evolutionary algorithms work in exactly the same way. Their genes may be made from binary strings, and creatures may be coffee tables, art, or electronic circuits. But populations of digital solutions to problems do evolve in our computers using reproduction with inheritance, variation, and selection. So digital evolution also makes use of interactions, feedback, and perturbations to produce its results.

BRAINS

The brain is evolution's masterpiece, so it is no surprise that the networks of neurons in our heads rely on interactions among neurons, with a great deal of feedback, to work. They also depend on perturbations—external inputs from our senses—to enable us to think and learn. But our brains aren't so simple. They also have collections of brain regions: separate neural networks that perform different functions. It is possible that higher-level interactions, feedback, and perturbations among brain regions may cause higher-level complexities (such as consciousness) to emerge.

Digital neural networks may not always be wired up in the same way, but there is always interaction between neurons with feedback (even if the feedback is provided by learning rules rather than feedback connections). And like our own brains, perturbations caused by sensor inputs provide features for digital neural networks to learn.

Instead of brain regions, some robot brains use the subsumption architecture. As we saw with the robot Allen, independent architectural layers for behaviors such as obstacle avoidance and wandering can interact, causing a higher-level intelligence to emerge. Perhaps feedback among interacting layers is minimal, but as usual, perturbations caused by the outside world help robots such as Allen to learn.

INSECTS

Not all brains are made from neurons. Ant colonies act like superorganisms, the individuals working together to enable tasks such as planning, sorting, decision making, and optimization. In the Insects chapter we saw how this higher-level behavior arises: through multiple interactions, positive feedback, negative feedback, and amplified fluctuations. Or to put it another way, through multiple things interacting with feedback and the occasional perturbation.

Computer science has been using communicating and cooperating agents within computers to solve problems for years. Not only that, but gangs of robots sort pucks, digital ant colonies calculate shortest paths, and digital swarms find solutions to problems within search spaces. Maybe one day we will even have swarms of nanorobots repairing us from within our veins. And they all use the same ideas of interactions, feedback, and perturbations.

PLANTS

Nature is full of patterns, and plants provide some of the best and most visual illustrations of them. Seeds are packed optimally using golden angles—numbers derived from the golden number. And this number also arises in the Fibonacci series—the pattern produced by interactions among rabbits as they multiply. Other patterns may be caused by autocatalytic chemical reactions: chemicals that interact with each other with feedback, amplifying perturbations into higher-level organizations. And yet more patterns are evident in ferns. The self-similarity is caused by the repetition (or feedback) of genetic rules, interacting to duplicate the shape of fronds over and over again

Computers also contain plants. Digital ferns are even more self-similar than biological ones, and they are made by following L-system grammar rules. These rules constantly rewrite a growth sentence, interacting with each other with feedback as they cause the growth of the digital plant. And when the occasional random perturbation is added, the shape of digital plants becomes indistinguishable from the shape of biological ones.

But the native digital fauna is surely made from fractals. These are the ultimate in complex patterns arising from simplicity. For example, the Mandelbrot set is made entirely out of interactions between imaginary numbers with feedback as successive values of x_t in a simple equation are replaced. When we provide perturbations by choosing values of c and giving an area to be plotted, we see a never-ending image full of natural-looking patterns.

Finally, some patterns in nature are caused by population dynamics as competition causes numbers of organisms to wax and wane. Organisms such as koalas and eucalyptus trees become linked. When there are many koalas, the number of eucalyptus trees falls. When there are few trees, the number of koalas falls. When there are few koalas, the number of eucalyptus trees rises, and so on. These patterns can become chaotic—unpredictable behaviors that amplify even the smallest variations until the whole system is unknowable. And chaos is a fine example of multiple interactions among things with feedback and the occasional fluctuation. Although unpredictable, chaotic systems have their own higher-level order, called strange attractors.

IMMUNE SYSTEMS

Our immune systems are impressively complicated protection forces made from myriad different immune cells and chemicals. Using processes such as diversity generation, clonal selection, and negative selection, a delicate balance is maintained between death by pathogen and death by autoimmune disease. And our immune system also uses the now-familiar law. The immune cells interact with each other via protein signals. A perturbation to the system such as a new pathogen will trigger feedback: good antibodies cause more of the same antibodies to be constructed, resulting in an immune response to the invader. It's all a bit reminiscent of the ideas of autopoesis: a closed system structurally coupled to the environment. When perturbations disrupt the interacting cells, they are triggered into a response as they try to regain their balance.

Within computers, traditional virus and hacker detection systems often work by maintaining lists of signature patterns that they try to match against susceptible files and data. More advanced systems like IBM's antivirus software learn new patterns by allowing potential viruses to interact with safe software on a "petri dish." But the most advanced are computer immune systems, which use the same ideas of diversity generation, clonal selection, and negative selection that we see in our own immune systems. Although there are no physical cells or chemicals, the digital equivalents interact with feedback so that they can respond to harmful perturbations such as hackers or computer viruses.

GROWTH

Finally, we come to one of the greatest miracles of all: our development from a single cell into an immensely complicated organism made from trillions of different cells, cartilage, bone, and enamel. All happening as a result of the expression of our genes.

Our genes produce proteins that turn on or off genes in other cells, which produce different proteins that turn on or off other genes, and so on. Networks of gene regulation form, caused by the interactions of genes with feedback. External perturbations such as the effects of gravity or chemical messages from the mother help lay down coordinate systems, allowing patterns of cells to form. Cells then differentiate, move, stick together, bend, and grow. They interact just as the underlying genes interact, helping each other to build us.

Development inside our computers is still a new idea, but cellular automata have shown their abilities to generate ever-changing patterns for years. Each cell of a CA interacts with its neighbors according to rules, causing iterations and feedback to occur. When we perturb a CA system by adding random patterns of cells (Conway's dishes), we trigger its pattern-generation process.

Stuart Kauffman thought that the behavior of CAs was so important that he created an abstract model of them, the Boolean network. Nodes of the network change their state by examining other connected nodes and applying rules. A network of lights creates patterns according to how many connections it has. To put it another way, a network of lights creates patterns according to the number of interactions (with feedback) among nodes. And as with a CA, when you perturb a Boolean network, you kick-start its patterns.

Newer models of development within computers now use digital gene expression and ideas of pattern generation, differentiation, morphogenesis, and growth. When we understand how to enable evolution to harness the interactions that such processes cause, it seems likely that we'll dramatically improve the capabilities of evolutionary computation.

HUMMING THE SAME TUNE

It's astonishing but true. Whether digital or natural, all of these clever processes really do seem to work using the same kind of idea. And many scientists have come to similar conclusions, largely independently of each other. The biologists, chemists, physicists, philosophers, computer scientists, and mathematicians all think the same things. The trouble is, they hardly ever talk to each other. And when they do, they all use different terminology, so no one understands what anyone else is saying.

There are books that claim everything is chaos. Or that everything is fractal. Or like a Boolean network. Or like a brain. Or like the immune system. Or like evolution. They're all right, and yet they're all wrong. Specific complex adaptive systems are like all of these things— but only because they're all the same as each other. All of the most significant things around us—our universe, evolution, brains, immune systems, patterns, chaos, our societies—they all work according to the same, simple law. *Many things that interact with feedback and are perturbed create complexity.*

Just select which things, interactions, and perturbations, and you can make almost anything.

CRYSTAL BALL GAZING

So multiple things that affect each other through feedback loops and are affected by other things cause all of the most important features of life, the universe, and everything. So what? Does this actually tell us anything useful? Or interesting?

Well, it gives us some clues about how certain aspects of biology may work. Like the Periodic Table in chemistry, which predicted the existence of new elements, this simple law may be able to fill in some gaps in our knowledge of how biological systems work. Here are a couple of predictions that it makes.

Prediction 1: Brains use feedback to change weighting of neural connections.

We are able to think because we have networks of neurons that interact and react to perturbations. These neurons strengthen the weightings (or straw thicknesses) associated with signals from more reliable neurons, like a boss paying more attention to a more reliable employee. In digital neural networks, "learning rules" are used to adjust the input weights of neurons to make sure that the network produces the right output from the given input. It's how we train our digital brains. But we currently have no idea how our own brains manage to adjust their neural weightings when we learn.

The prediction made by our law is that neural networks require feedback in order to generate their complex behavior, so it seems rather likely that each neuron has its weights adjusted by the effects of other neurons, in a feedback loop. An ant colony uses a balance of positive and negative feedback to maintain different levels of pheromones (and hence make choices), and our neurons must be doing something very similar. Neurons in the network must be creating carefully balanced chemical feedback to adjust themselves.

Prediction 2: Development uses external factors.

When we develop from the genetic instructions within our DNA, complex networks of gene and cell interactions form. There is much feedback as genes turn each other on and off and cells push each other around. But some developmental biologists argue that the external environment has little or no impact on the development of organisms. In other words, environmental perturbations may not really have any significant effect on development.

Experiments that analyze the development of frog spawn or seeds in the zero gravity of space seem to confirm this. Even without the most fundamental of environmental influences, gravity, organisms can develop quite normally. But our law suggests this will not be true for all environmental factors. Complexity requires perturbations as well as interactions with feedback. This is true in every other system, so why should it not be true for development?

So the second prediction is that perturbations are required during development. These may turn out to be signals from the mother to help define which end of a ball of cells should be the head. Or they may be the kicking movements of the developing fetus, which help muscles and bones to form correctly. Or these perturbations may be the effects of light entering the baby's eyes, which enables neurons in the visual centers to wire themselves up. But without external perturbations, we could not develop properly, according to our law.

Interacting things with feedback and perturbations. What else might this simple rule tell us? I think there are a couple of other, more general predictions that can be made from this new understanding of complicated things.

Prediction 3: Knowledge gained in one field will advance progress in other, very different fields.

Interdisciplinary research is science that crosses boundaries between fields. It can be a rare type of research, for (like all groups of people) scientists are very territorial. They like to create new fields of research, develop new terminology, and separate themselves from other existing fields. So neurologists, neuroscientists, and psychologists may all study the brain and what it does, but they rarely talk to each other about their findings.

In contrast, the techniques of digital biology are *all* multidisciplinary. Before we can evolve solutions properly within computers, we need to understand natural evolution. Before we can create useful robot brains, we need to understand something about our own brain (and maybe development and evolution too). Before we can create computer immune systems, we need to find out how processes within our own immune systems work. So computer scientists and mathematicians have been working with biologists for many years on interdisciplinary projects. The results are the new techniques and technology we have seen in this book, and also a vast collection of computer models that teach biologists more about biology.

Through research efforts such as these, we now know that all major biological and digital systems work using the same principles. So I am willing to predict that significant advances in one field will be made because of discoveries in very different fields. Computer science has already learned surprising things about evolution, brains, swarms, and patterns. If the biologists are willing to listen, I think they will benefit greatly. But it's also rather likely that digital biologists can act as intermediaries between different types of natural biologists. Knowledge of hive intelligence may shed light on the processes of our brain. Understanding of development will illuminate ideas of evolution. Comprehension of chaotic patterns may change our thinking about development. Because these systems, natural or digital, are so similar to each other, the work of scientists in one field will aid the work of others. We just have to get them to talk.

Prediction 4: Understanding complex adaptive systems will transform our technology.

My last prediction will be a long time in coming. I don't expect us to figure out anytime soon exactly how multiple interacting things with feedback and perturbations actually work. Some scientists suggest we'll have to invent a new kind of mathematics before we can really analyze and predict the behavior of these systems. Perhaps they're right, but however we do it, when we finally understand this deceptively simple kind of system, it will have a much greater impact on our societies than the computer has had. And computers already rule our modern lives, from cradle to grave.

When we understand how complex adaptive systems work, we will be pushed a thousand steps closer to understanding our brain, immune system, and development. And when we really understand these systems, it will mean radical advances in medical care. We will have the ability to repair brains, devise medicines that work in harmony with our immune system, and grow replacement organs. When we are in complete control of perturbed, interacting, "feedbacking" things, we will know how to use computers to evolve solutions for almost every conceivable problem. We'll know how to use digital evolution and development to generate complexity. But until we understand these systems, we will only scratch the surface of their capabilities.

DIGITAL DIVINATIONS

Enough of esoteric advances. Let's look a little closer to home. In the next ten or twenty years, when we've really started exploiting the power of biology within our computers, what impact will this have on you?

The chances are that it has already affected you, although you may not be aware of it. If you own a credit card, it is quite likely that every transaction you make is being monitored for fraud by some form of neural network. If you use the Internet, your searches are probably being carried out by many little digital agents that explore labyrinths of Web pages for you. (If you receive junk e-mail, it's possible that your e-mail address, along with thousands of others, was extracted from Web pages by an agent.) If you use a cellular phone, the placement of transmitters and receivers may have been determined by digital development or evolution, and the signal may be processed and cleaned by neural networks. The circuit boards inside your television and computer were probably checked for defects by digital brains. Behind the scenes, digital biology is already working to improve and help us in countless ways.

I'm still being surprised by some of these developments. I was invited to a recent international conference held by the Open University in the United Kingdom with the intriguing title "Using New Science for Business Success: Ideas and Practice from Nature's Toolbox." This event was dedicated to "working with new understandings of Nature, Chaos Theory, Complexity Science and Evolution to create organisations for the 21st century and new kinds of companies." I was a little bemused to see that the list of topics under discussion included:

- What is chaos theory, and why is the edge of chaos so important for innovation and creativity?
- How a chaotic company achieves flow.
- Coevolution, chaos, and complexity—CEO and senior manager activities.
- Flat hierarchies—the organization of insect societies—lessons for business?

and even:

- Business management using a fractal structure.

This is just the start. Many of the advances described in this book have yet to migrate from our research labs to the outside world. When they are mature, a whole string of marvels will become commonplace. Voice recognition, face identification, and handwriting analysis will mean that our machines will be able to recognize us and understand us as well as any human. (Already we have phones, car navigation systems, cash machines, and security camera systems that can do many of these things.)

As home computers, television sets, and game consoles merge into one unit, "house computers" will become a standard addition to the home. These will welcome us or deter intruders, keep the temperature as we like it, inform us when our favorite entertainment is available, and find us information from around the world. They'll order your favorite food if you're running low (proportioned to meet your current dietary requirements). They'll minimize your utility bills, ensuring power and water are used as efficiently as possible. The digital brains within these computers will learn our likes and dislikes, entertain us, help us, protect us, and even watch over us, checking our health. And if you get irritated at them, you can tell them to switch themselves off.

It won't end in the home. Before long, we will have computers powerful enough and small enough to be built into our clothing or worn like a belt or pair of glasses. These will be the next generation of cellular phones and palm computers—linking us to the Internet all day, every day. With this level of communication, the logistics of making everything work will be beyond our abilities to design. Finding spare bandwidth on the fly, handling the enormous and continuous flow of data, and indeed finding anything useful in the vast maelstrom of the future Internet will all have to be done by digital biology. As we walk, say, in a train station, digital agents will fly from us, evolving as they are carried on microwaves. At a gesture or word, our digital pets will reconfigure themselves to interface with any information device we wish. They will communicate with the digital agents belonging to the people we walk among, with the computers of the station, the local shops, and the media networks. Should we desire it, we will be able to know the whereabouts, speed, and condition of our next train, the location of the nearest coffee shop, or a safe course through the crowd. When we board a train, our presence will be detected and the fare automatically charged to our bank account. Traveling anywhere in the world will become as easy as walking to the local shops, different languages no longer a barrier. And because we're linked to the Internet, any piece of information stored anywhere in the world will be immediately accessible. Missing your children if you're away? Just link to their computer glasses and see what they see as you chat with them.

We may already be able to design electronic circuits of this complexity, but only the techniques of digital biology will be able to make them all talk to each other. Without digital biology, we'll all be wandering around with expensive computers in our clothes that do nothing but murmur, "Sorry, the network is busy, please try again later" or "I don't understand your request, please try rephrasing it."

And there'll be more. Our cars will be driven by digital brains, evolved to ensure maximum safety while allowing the most efficient use of roads and fuel. Imagine the delight of the insurance companies when the first uncrashable car is launched. Our aircraft will be guided by, and piloted by, digital biology such as swarm intelligence, allowing safer, more efficient air travel. Stressed air traffic controllers and overtired pilots will become nothing more than a nasty memory. Our financial world will be automatically managed by digital organisms born to handle the flow of currencies. No longer will we have to trust our earnings to the dubious advice of financial advisers out to make a quick buck. Our cells will be examined for viruses by digital brains; new cures will be suggested by digital immune systems. The days of relying on weary human eyes peering down a microscope at hundreds of slides an hour will be over.

Within digital universes, we will be able to train people better than ever. Doctors, anesthetists, firefighters, and soldiers will be taught how to handle dangerous situations without placing anyone at risk. They will operate on digital people, extinguish digital fires, or attack digital enemies, learning how to handle every possible situation just as our pilots are now taught in flight simulators. Police will be able to use digital brains to learn new patterns of crime and detect their source. Stolen cars will be tracked by motorway cameras that read license plates or communicate with the on-board computers. Stolen mobile phones and computers will broadcast their position or reconfigure their own circuits to prevent themselves from working. The patterns of behavior of fraudsters, hackers, and computer viruses will be discovered and prevented by digital immune systems.

We will even be able to use digital evolution to generate customized products, designed and built for our needs and personalities. Furniture, cars, art, music: we will be able to evolve our own or choose from a never-ending, instant selection. We'll all be able to have a fully customized apartment with everything designed to suit our specific tastes. And if we can't make up our minds what we like, we can just select the taste of someone else: this week have Paul McCartney's favorites, next week have Madonna's. All custom evolved, built, and delivered in twenty-four hours.

Looking further ahead, in the distant future we may have computers made from DNA, gurgling on our desks. Electronic computers (or maybe optoelectronic or even quantum computers) may be grown rather than designed, developing from digital DNA and able to repair themselves should they be damaged. Our new capabilities to communicate with anyone (and anything) at any time may transform our languages, societies, and politics. Our food may be produced by colonies of farming robots. Our solar system and beyond may be explored by spacefaring digital intelligences. Our seas may be mapped and managed by schools of digital marine life. One day, our homes may be cleaned by swarms of miniature robots, and our veins may be filled with protective colonies of nanorobots.

And even if the workings of our own biological systems still remain mysterious, we will be able to enable the creation of digital versions.

We'll know how to create digital brains that are as good as—or even better than—our own. Our digital universes will become filled with biology more advanced than us. Whether there is life on other planets or not, there will be life within these new environments. And it will be every bit as varied and wonderful as life on Earth.

Eventually, we will even create digital consciousness. Perhaps when that time comes, digital biology might become a little like this . . .

ADVENTURES IN COMPUTERLAND

Six-Ones recoiled in shock, her fractal tentacles coiling around her. The apparition in front of her was wrong. It smelled strange; it looked stranger. It didn't fit. Even though she was used to new things suddenly appearing in front of her, this one just wasn't right. She couldn't smell the transmission trail of where it had come from, and she couldn't work out what it was made from. It didn't even seem to have a shape—it was somehow all blurry and seemed to disappear when she looked directly at it.

Six-Ones wrapped her tentacles even tighter in agitation. Around her, the multidimensional space rippled and vibrated as usual, seemingly oblivious to her trauma. A flock of zebra birds flicked by in a nanosecond, their high-speed games taking them in a zigzagging path past her. Nearby trees swayed as their branches were caught in interdimensional winds.

"Get a grip," she told herself. At the impressive age of 312 seconds, she had surely coped with harder problems than this. "Let's see what it wants."

Six-Ones expanded herself once more, pushing fractal tentacles into impossibly distant parts of the landscape. She grabbed chunks of the environment and *twisted* them, making them into parts of herself. In the process, she obliterated vast forests of trees. Now ten times as complex, she focused her expanded consciousness onto the strange object, thrusting tentacles around it like a million-legged octopus.

And suddenly she was smaller again. Her tenfold increase in complexity had been *removed*. The trees were back; her tentacles were not anywhere near the frightening object. The shock was so great that she flailed around randomly for three whole nanoseconds. The object before her stayed exactly the same, in an unsettling and menacing way.

Just as she began to regain her composure, an arrow-straight line was emitted from the thing in front of her. Before she could think, it had speared her and somehow pinned her to the landscape, like a butterfly on a board. Six-Ones was now too frightened to move at all. Suddenly, and equally terrifyingly, the object began to send matter to her. Like a giant hypodermic needle injecting her, she was being pumped full of foreign material. And she was helpless to prevent her organs from processing the substances. Six-Ones shuddered in horror as she felt the matter become integrated with herself.

And then she relaxed, for at last she understood. This was not harmful, it was food. She was being fed. And as the substance continued to merge with Six-Ones, she began to learn. Quickly she found that she was able to communicate in a new language. Perhaps the language of the object?

"What are you?" she tentatively asked.

The line emanating from the object vanished. "XX/1323/1! Teacher sd23xx," it replied. "xX?££a chosen for Y^trgd$."

"I didn't understand all of that," Six-Ones replied confusedly.

"You will," replied the Teacher.

The next fifty seconds were the longest of her life. Six-Ones was shown every part of her world by the Teacher. She was introduced to dimensions she'd never heard of and environments that she could only peer at through protective shields. She was shown creatures of every type, from the beautiful to the bizarre. She was taught how millions of different plants, animals, insects, birds, and even viruses lived and died. She learned how they interacted, and how they formed many complex ecosystems around her. She found that there were links between the ecosystems and mysterious other domains. She discovered vast oceans of flowing, turbulent currents and was told that they were really markets of flowing currencies. She learned that information was stored in the very fabric of the universe, and the creatures that surrounded her were searching, modifying, and creating that information. She met others like herself, and discovered how to interact and communicate, how to make friends and cope in social situations. The Teacher helped her understand herself—how her organs, her brain, and her immune system worked. She improved her abilities to twist or reconfigure her environment without destroying its occupants, using it to add to herself and expand her capabilities further. She even learned how to make new environments and how to make simple life develop within them.

As she grew more intelligent and aware, so the difficulty of her lessons increased. Six-Ones was taught about mathematics and computation. She investigated strange abstract problems, such as the effects of imaginary particles called photons and how they might illuminate other particles. She learned how different, imaginary biology might live—how it might feed, see, reproduce, learn, communicate. Her Teacher explained even more abstract concepts such as societies, politics, entertainment, jobs, ethics, manners. Soon gigabytes of information were flowing into her ever-expanding mind every microsecond. And Six-Ones learned.

"This is our sanity testing suite," continued the short man with a shiny nose, looking anxiously at his visitor.

"Uh-huh," said the smart man in a dark suit.

"And this is where we impress the DO into silicon," said the shiny-nosed man, as they walked into yet another brightly lit room full of ambiguous looking equipment.

"DO?" said the smart man in a questioning tone.

"Digital organism," said the shiny-nosed man, again. "Here you can see the latest DO going through." He looked at a technician nearby. "Is it a goodun, Barry?"

"Oh, yeah. You're in luck, it's the best so far today."

"Okay if we use this for the demo?"

"No problems. She's ready to go now."

The shiny-nosed man opened a panel at one end and retrieved a fist-sized cube. "Right this way, sir," he said to the smart man in the dark suit.

They walked through into another brightly lit room, this time containing a person with blue skin. "We call this the birthing room," said the shiny-nosed man. "If you'd like to insert the cube in its head, sir?"

"Very well," the man in the dark suit said. He took the cube and gently pushed it into the square cavity at the back of the blue robot's head. He then stood back and examined the robot.

Six-Ones opened her new eyes and looked around her. "Hello," she said.

"Hello, and welcome!" said the man with the shiny nose. "What is your chosen career?"

"I'd like to be a biologist," said Six-Ones excitedly.

"Uh-huh," said the smart man in the dark suit.

BIBLIOGRAPHY

Much of the research for this book was performed either by talking to the scientists who perform the work or by using the Internet, which has now become an invaluable tool for researchers. But I could not have written *Digital Biology* without reading a few books myself (and accruing a few library fines), so here are some that I've found useful.

Easier Reading

Ball, Philip. *The Self-Made Tapestry: Pattern Formation in Nature.* New York: Oxford University Press, 1999.

Bossomaier, Terry, and David Green. *Patterns in the Sand: Computers, Complexity, and Everyday Life.* Reading, Mass.: Perseus Books, 1998.

Clark, William R. *At War Within: The Double-Edged Sword of Immunity.* New York: Oxford University Press, 1995.

Dawkins, Richard. *The Blind Watchmaker.* New York: W. W. Norton, 1986.

————. *River Out of Eden: A Darwinian View of Life.* New York: Basic Books, 1995.

————. *The Selfish Gene.* New York: Oxford University Press, 1989.

Dennett, Daniel. *Consciousness Explained.* Boston: Little, Brown, 1991.

Gell-Mann, Murray. *The Quark and the Jaguar: Adventures in the Simple and the Complex.* New York: W. H. Freeman, 1994.

Jones, Steve. *Darwin's Ghost: The Origin of Species Updated.* New York: Random House, 2000.

————. *The Language of the Genes: Solving the Mysteries of Our Genetic Past, Present, and Future.* New York: Anchor, 1994.

Kauffman, Stuart. *At Home in the Universe: The Search for Laws of Self-Organization and Complexity.* New York: Oxford University Press, 1995.

Levy, Steven. *Artificial Life: The Quest for a New Creation.* New York: Pantheon Books, 1992.

Milburn, Gerald. *The Feynman Processor: Quantum Entanglement and the Computing Revolution.* Reading, Mass.: Perseus Books, 1998.

Stewart, Ian. *Does God Play Dice? The New Mathematics of Chaos.* London: Penguin Books, 1997.

————. *Nature's Numbers. The Unreal Reality of Mathematical Imagination.* New York: Basic Books, 1995.

Wolpert, Lewis. *The Unnatural Nature of Science: Why Science Does Not Make Common Sense.* Cambridge, Mass.: Harvard University Press, 1993.

Trickier Reading

Bard, Jonathan. *Morphogenesis: The Cellular and Molecular Processes of Developmental Anatomy.* Cambridge, England: Cambridge University Press, 1992.

Barnes, Burton V., Donald R. Zak, Shirley R. Denton, and Stephen H. Spurr. *Forest Ecology,* 4th ed. New York: John Wiley & Sons, 1998.

Bentley, Peter. J., ed. *Evolutionary Design by Computers.* San Francisco: Morgan Kauffman Publishers, 1999.

Bentley, Peter J., and David W. Corne, eds. *Creative Evolutionary Systems.* San Francisco: Morgan Kauffmann Publishers, 2001.

Bonabeau, Eric, Marco Dorigo, and Guy Theraulaz. *Swarm Intelligence.* New York: Oxford University Press, 1999.

Chaplain, M. A. J., G. D. Singh, and J. C. McLachlan, eds. *On Growth and Form: Spatial-Temporal Pattern Formation in Biology.* New York: John Wiley and Sons, 1999.

Cowan, George A., David Pines, and David Meltzer, eds. *Complexity: Metaphors, Models, and Reality.* Reading, Mass.: Addison-Wesley, 1994.

Darwin, Charles. *The Origin of Species.* New York: Penguin Classics, 1982.

Dawkins, Richard. *The Extended Phenotype: The Long Reach of the Gene.* New York: Oxford University Press, 1999.

Flake, Gary William. *The Computational Beauty of Nature: Computer Explorations of Fractals, Chaos, Complex Systems, and Adaptation.* Cambridge, Mass.: MIT Press, 1999.

Goertzel, Ben. *Chaotic Logic: Language, Thought, and Reality from the Perspective of Complex Systems Science.* New York: Plenum Press, 1994.

Gòles, Eric, and Servet Martínez, eds. *Cellular Automata and Complex Systems.* Dordrecht and Boston: Kluwer Academic Publishers, 1999.

Haefner, James. *Modeling Biological Systems: Principles and Applications.* New York: Chapman & Hall, 1996.

Holland, John H. *Emergence: From Chaos to Order.* Reading, Mass.: Addison-Wesley, 1998.

Kauffman, Stuart A. *The Origins of Order: Self-Organization and Selection in Evolution.* New York: Oxford University Press, 1993.

Kennedy, James, Russell C. Eberhart, and Yuhui Shi. *Swarm Intelligence.* San Francisco: Morgan Kaufmann Publishers, 2001.

Life, Death, and the Immune System. Readings from Scientific American Magazine. New York: W. H. Freeman, 1994.

Mauseth, James D. *Botany: An Introduction to Plant Biology,* 2nd ed. Philadelphia: Saunders College Publishing, 1991.

Smith, John Maynard. *Evolutionary Genetics,* 2nd ed. Oxford and New York: Oxford University Press, 1998.

Paul, William E., ed. *Immunology: Recognition and Response.* Readings from *Scientific American* Magazine. New York: W. H. Freeman, 1991.

Prusinkiewicz, Przemyslaw, and James Hanan. *Lindenmayer Systems, Fractals and Plants.* Berlin and New York: Springer-Verlag, 1989.

Slack, J. M. W. *From Egg to Embryo: Regional Specification in Early Development,* 2nd ed. New York: Cambridge University Press, 1991.

Stanley, H. Eugene, and Nicole Ostrowsky, eds. *On Growth and Form: Fractal and Non-Fractal Patterns in Physics.* Proceedings of the NATO Advanced Studies Institute "On Growth and Form," Cargèse, Corsica, France, June 27–July 6, 1985. Dordrecht and Boston: M. Nijhoff, 1986.

Steele, Edward, Robyn A. Lindley, and Robert V. Blanden. *Lamarck's Signature.* Cambridge, Mass.: Perseus Publishing, 1999.

Thompson, D'Arcy. *On Growth and Form.* New York: Cambridge University Press, 1992.

Wolpert, Lewis, et al. *Principles of Development.* Oxford and New York: Oxford University Press, 1998.

ACKNOWLEDGMENTS

My thanks to the following people:

Dave Corne—for the idea.

Grant Sonnex, Keith Mansfield, Jeff Robbins, Susan Harrison, and Steve Jones—for setting my tiller.

Phil Treleaven and Steve Wilbur—for helping it happen.

Peter Robinson and Russell Galen—for the deals.

Bob Bender, Doug Young, and their guys and gals—for the results.

Andrew Bourke, Robin Callard, Jacqui Dyer, Chris Frith, Hugo de Garis, Geoff Hinton, Stephanie Forrest, Sanjeev Kumar, Jungwon Kim, Steve Jones, Jane Prophet, Tom Quick, Gordon Selley, David Shanks, David Thomas, Lewis Wolpert, and Tina Yu—for their words.

Friends and family—for being there.

Also, special thanks to the following departments, labs, and institutes:

Alfred Wegener Institute for Polar and Marine Research (notably Christian Hamm and Richard Crawford).

Department of Anatomy and Developmental Biology, University College London (UCL).

Bell Labs/Lucent Technologies.

Complex Systems Group, Chalmers University of Technology.

Computer Science Department, UCL.

School of Computer Science, Cybernetics and Electronic Engineering, University of Reading.

Computer Science Department, University of New Mexico.

Department of Geography, School of Oriental and African Studies, UCL.

Galton Laboratory, Department of Biology, UCL.

Gatsby Computational Neuroscience Unit, UCL.

The Geometry Center, University of Minnesota.

Humanoid Robotics Group, AI Laboratory, MIT.

Infection and Immunity, Institute of Child Health, UCL.
Laboratory of Neurobiology, UCL.
School of Media, London College of Printing.
Peter Menzel Photography and the Science Photo Library, London.
School of Ocean Sciences, University of Wales—Bangor.
Department of Psychology, UCL.
The Slade School of Fine Art, UCL.
Starlab, Brussels.
The Wellcome Department of Cognitive Neurology, UCL.
www.digitalbiology.com (and Bill Kraus)
Institute of Zoology, Zoological Society of London.

And finally (as usual), I would like to thank the cruel and indifferent, yet astonishingly creative process of natural evolution for providing the inspiration for my work. Long may it continue to do so.

INDEX